For Susan & Robby—
hope you enjoy
this step back in
time.
Love,
Helen

DADDY'S DIARY

The War Journal of B-17 Pilot Lt. Edwin "Snake" Walkup

by
Helen Walkup Cairns

Written by Helen Walkup Cairns
Based upon the World War II journal of her father,
Lt. Edwin "Snake" Walkup

Daddy's Diary:
The War Journal of B-17 Pilot Lt. Edwin "Snake" Walkup

ISBN: 978-1-956793-04-8
First Edition

With love, gratitude, and devotion,
I dedicate this book to my parents
Mary Elizabeth and Edwin Fenby Walkup.
Having them love and guide me was a blessing,
and I miss them every day.

"Without the B-17,
we may have lost the war."

— General Carl Spaatz
World War II American Air Commander in Europe

The Lady Satan from the 452nd Bomb Group. After 85 missions, this ship was lost to antiaircraft fire on February 6, 1945. Photo from the Irwin Math Collection [Acc. # 2008.0151] / National Museum of the Mighty Eighth Air Force, Pooler, Georgia.

IN APPRECIATION

First, I thank my father for faithfully writing in his diary every day, even on days when he had been in the cockpit for over eight hours and had faced being hit by flak from the ground or bullets from German fighters. And thanks also to my mother, who helped me with the transcription of Daddy's diary.

In addition, I owe a huge debt of gratitude to the following:

Kyle Miller for referring me to my editor, Jim Huth of WordEquity. Without him, you would not be reading this book. He led me with knowledge and grace down the path to finally being able to publish.

Chuck Ickes, Major General (retired), U.S. Air Force, for reading the very rough draft of the book and for making corrections and important suggestions, and for being a great cheerleader.

Bill Frederick, former mayor of Orlando, Florida (1980-1992), for reading the rough draft and, like Chuck, making valuable corrections and suggestions. Bill also was encouraging and knowledgeable as he published a book which you may want to read: "Just Jeeps and Such."

Dr. Vivian Rogers-Price, Research Center Director, National Museum of the Mighty Eight Air Force, for answering my questions and providing me with additional research materials. How fortunate I was to have her expansive knowledge available.

To each of these generous people, I give thanks. For anyone who has not visited the WWII museum in New Orleans, I highly recommend it. While there, be sure and see the WWII jeep collection donated by Bill Frederick. And worth a visit is the National Museum of the Mighty Eighth Air Force, located in Pooler, GA, just north of Savannah. There you can see a restored B-17 and other informative and moving exhibits.

And finally, thank you to each of you for reading this diary written by my father, who celebrated his 29th birthday while fighting for the freedom of our country, our parents, each of us, and our children and grandchildren. We have been blessed.

TABLE OF CONTENTS

FOREWORD

by Charles V. Ickes II
Major General (retired)
United States Air Force

During World War II, the Eighth Army Air Force was tasked with flying bomber missions out of Great Britain. The B-17 Flying Fortress was the workhorse of this campaign and often saw some of the fiercest fighting. These incredible aircraft were flown by men from all parts of America. Men who hailed from large, medium, and small towns and, for the most part, were inexperienced aviation novices.

Yet, these young aviators turned out to be men of extraordinary courage who would significantly impact the war in Europe and play a major role in bringing Germany to its knees.

They were tasked with flying their war machines into some of the most heavily fortified airspaces in the world. They flew in large, static formations and were tasked with daylight precision bombing. The German Luftwaffe attacked them relentlessly on every mission and the skies were filled with antiaircraft fire. The missions were long and hazardous, but were critical to the war effort.

Lt. Edwin "Snake" Walkup was one of those incredible Americans — more than 16 million who served. Snake came from the small town of McIntosh, a short distance outside of Gainesville, Florida. He was a college graduate, highly educated, and a superb athlete. Snake kept a diary of his time flying the B-17 during 1944. His writing was clear, direct, and succinct. He covered his experiences, both flying and nonflying, and took a candid assessment of his surroundings.

In 2020 I met Lt. Walkup's daughter Helen Cairns. During our initial encounter, Helen asked about my time serving in the United States Air Force. I shared with her that during my nearly 38-year career, I was blessed to have flown some of the greatest fighter aircraft the world has ever known, including the F106, F16, and F15.

Helen proudly revealed to me the story about her father. Her pride and passion were infectious, and I asked her if she would share her father's diary with me. I then learned she was well on her way to turning it into a book to share with anyone who might be interested. I was honored when she asked if I would read her transcription and comment on it.

I found Snake's diary to be highly informative and

entertaining. He had a great sense of humor, and like most of the junior aviators I knew, he had a belief that he knew how to do it better. He pulled no punches and captured the essence of what it was to be in the midst of war as a B-17 Bomber pilot.

I am the son of a decorated WWII and Korean War Army infantry veteran. My respect for my father and all those who served during those years is immeasurable. They were average men asked to do extraordinary things. They stepped up to serve when the nation called, and most served proudly and honorably.

Snake was one of these heroes, and my only regret is that I never met him in person. However, getting to read his diary was a special treat and one that I will always cherish.

Charles V. Ickes II
Major General (retired)
United States Air Force

PROLOGUE

This Is My Father

Before reading this diary, you may like to know a little about the young man who wrote it and how it came to be printed.

Edwin "Snake" Walkup was born in 1915 in a little town in north-central Florida, McIntosh: population then and now of about 300. He lived most of his life there. He was the second of six siblings; four brothers and two sisters. His father, Sam, was a farmer of cotton, strawberries, watermelons, and citrus. His mother, Elizabeth, was a homemaker who had a garden, raised chickens, canned, sewed, and did the things that ladies of that era did. In the white clapboard, two-story house where they lived, the boys shared a tiny bedroom while the girls had a large bedroom, complete with a fireplace and closet. The boys each had a box, which they stored under their beds. Each box contained all of their clothes.

Grade school was in a small community school. For one of those grade school years, Daddy was bedridden with rheumatic fever. During those twelve months, he lived in the hospital at Johns Hopkins where his great-uncle and the person for whom he was named, Dr. Edwin Fenby, practiced. Ninth, tenth and eleventh grades were in a somewhat larger school, just down the road from McIntosh in another small town. For twelfth grade, he went to the largest high school in the area, in Ocala, eighteen miles away.

While at Stetson University, Daddy (standing, right) was a waiter and short-order cook.

From there, he attended the University of Florida for two years and then Stetson University for three. While at Stetson, he played baseball, football, and ran track, all while working his way through college waiting tables at school banquets and being a short-order cook on the breakfast shift. He graduated with a degree in economics and business administration.

When he was drafted into the army in April of 1942, he was sent to Ft. Sill, Oklahoma, to work on geodetic computation. After marching in the heat and looking up to see airplanes overhead, he immediately applied for the U.S. Army Air Corps cadet training program. He was accepted in July 1942, receiving his wings in July 1943.

His pre-flight training was at Santa Ana, California. From there, he went to Thunderbird Field near Phoenix, Arizona for primary flight training. Basic flight training was at Pecos, Texas, and advanced training was at Ft. Sumner near Clovis, New Mexico. He graduated as a 2nd Lieutenant Army Air pilot assigned to B-17 bombers and got his wings there on July 27, 1943. He got a pass to go home and was married to his long-time sweetheart, Mary Elizabeth Harrison, in the Presbyterian church in McIntosh on August 1, 1943.

Accompanied by Mother, he was assigned for short periods of time at five different bases in Washington, Oregon, Iowa, and Nebraska until he departed for England on May 20, 1944.

Daddy (standing, center) with his B-17 crew.

After he came home from the war, he was in the Army Air / U.S. Air Force for an additional eleven years, during which time he was promoted to Captain. In an interview years later, he said that he and Mother were moved every few months to various bases in Florida. He returned to Europe, serving at Templehof Airport during the Berlin Airlift in the group that coordinated the transfer of the imported goods to their destination in West Berlin. (The Berlin Airlift was called Operation Vittles and was underway from June 1948 through September 1949.) When he returned from Berlin, we were stationed at MacDill Air Force Base in Tampa. He was off active duty in 1950 and discharged as a Captain in 1955.

Mother and Daddy before he left for the war.

Mother and Daddy had two children: my brother Edwin Jr. and me. He was the best father anyone could ask for, always loving us, encouraging us, and supporting us. He made a living running or owning lumber companies in or near Gainesville, Florida. During that time, he was always community-minded, serving on the McIntosh town council for twenty-seven years.

In the McIntosh Community Park, there is a plaque honoring Daddy and his devotion to the city.

He passed away November 28, 1996 at eighty-one years of age, having led a life that he could be proud of, and so could we. After he was gone, my mother found this diary. I promised her that I would transcribe it. Thanks to her for helping me with Daddy's handwriting, interpreting some of the words, and clarifying various things. After a short illness, Mother passed away on

March 2, 2015, just a few days shy of her ninety-fifth birthday. She had a fantastic memory and recalled a lot about the people and places mentioned in this memoir.

I hope you enjoy this glimpse into another time.

Helen Walkup Cairns

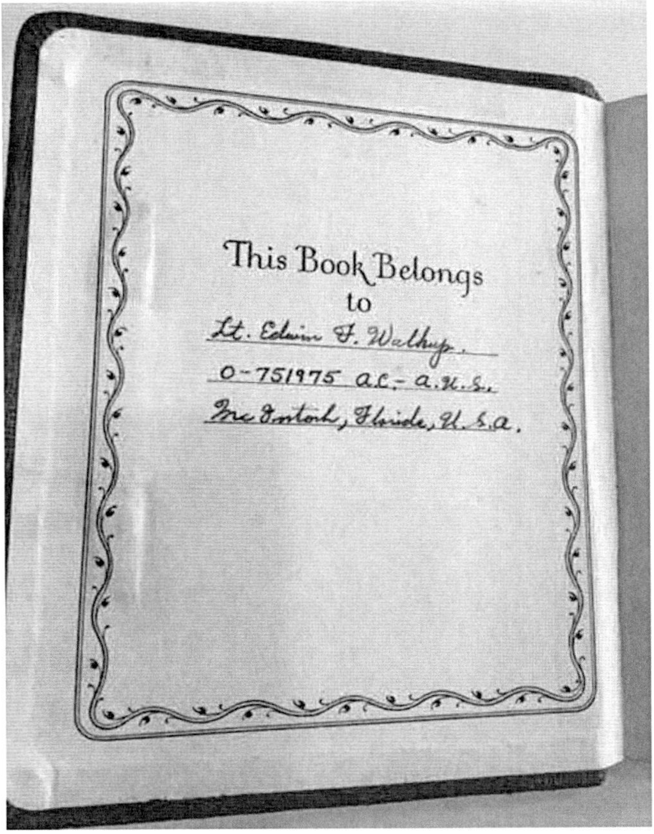

The inscription on Daddy's journal reads:
This Book Belongs to Lt. Edwin F. Walkup,
0-751975 AC-AUS, McIntosh, Florida , USA.

1

SOMETHING BREWING

Friday, May 19th, 1944
Kearney, Nebraska

Were supposed to take off for Presque Isle, Maine, at 0100 this morning but the flight was canceled. Played poker from about 0300 to 0700 and lost about $25. Then I went to bed and slept until noon. Called Mr. Detwielder and tried to buy a case of whiskey or scotch but didn't get it, for he was sold out.

Went to a show to pass the afternoon. Ate a late chow and went down to the ship* to preflight it and attempt another takeoff.

*Daddy's ship was the B-17. For more information on the B-17, please turn to page 203. B-17s are what Daddy flew the entire time he was in the European Theater of Operations.

Saturday, May 20th
Logged 9:30
4:00 Instrument, 5:30 Day CX

Took off at 0215 from Kearney, Nebraska, with a clearance to Presque Isle, Maine, but our destination was changed inflight and we landed at Lanier Field, Manchester, Massachusetts. We rode through a front with lots of thunderheads and turbulence, and due to static electricity, we had quite a lot of St. Elmo's fire on our props and wings.

The country in Indiana, Ohio, Pennsylvania, and New York with the new crops, nice and green, was beautiful. Went over to look at Niagara Falls but a fighter came up before we could see it well. We left and took up course. Saw the famed Erie Canal and the upper Hudson. Massachusetts was typical summer vacation country with green mountains and lots of lakes and streams.

Sunday, May 21st
Logged 6:30
All Day CX

Slept soundly last night and the cool morning felt good. Left Lanier Field at 0930 and landed at Gander Field, Newfoundland, at 1600. The flight was over heavily wooded country, which was sparsely settled but covered with lakes. We left the U.S. at 1136 and left the North American continent at 1230. It looked good as I took my last look for quite some time I suppose.

Newfoundland has more lakes than any place I have ever seen, there was lots of snow still on the ground and there was nothing but spruce trees. Gander has nice large runways and is used by both the British and us. The only signs of life I saw on the island were the spruce logs in the lakes being floated to the coast where they were loaded onto steamers for shipment.

Monday, May 22nd

Spent the day at Gander Field, finally leaving at 2030 GST*. It was rather cold and a leather jacket and gloves felt good. An American dollar here is worth $1.10 in Newfoundland money. They have some civilian girls working on the base and there is a guard around their barracks 24 hours a day. Heard the boys say the native girls thought nothing of shacking up on one end of the hall while their mothers prepared supper on the other end. Mixed drinks here sold for 15 cents and there were plenty.

Saw lots of British soldiers and planes at this base. Had our briefing for the overwater hop and weather briefing. We must fly at 17,000 ft. the first four hours and then descend. Going out I saw my first icebergs floating around the Atlantic like big ships.

*Greenwich Standard Time.

Tuesday, May 23rd
Logged 11:30
4:00 Night, 7:33 Day,
6:00 Instrument

Sun rose at 4:00 this morning as we were out over the Atlantic above an overcast. Went below the overcast and saw lots of merchant ships. Hit Ireland 20 miles south of course. Landed at Valley Airdrome near Hollyhead in Wales at 0745 to touch the UK for my first time. Sent Mary Liz a cable but couldn't say much other than safe and well. Changed all my U.S. dollars into British pounds and other queer-looking money.

Tried my first cup of English tea and I don't think I like it so well. Lorry is the English word for a truck. This left-hand traffic kills me — and so many bikes with no sort of brakes. Went to a WARF dance tonight; the girls here dance as we do and the music is the same. Did a folk dance called the Hokey Pokey, which reminded me of a square dance.

Wednesday, May 24th

Boarded the train at Hollyhead and departed at Stone in England five hours later. The ride, my first on an English train, was very good; the coaches were rather dirty from soot but there were no jerks and the rails were smooth. There are bells on these trains and the whistles seldom blow. The freight cars are small things not as large as our four-wheel trucks and they have only four wheels.

Saw some old castles and several barge canals on which horses were pulling the boats from the shore. Also saw the results of some of the German bombing, which was centered on gas and electric plants. Stone is in the center of the Lowlands. Rock fences are not so pronounced here and the fields are larger. The only thing seems to be pastureland.

Thursday, May 25th

Found out today that everything in the PX* is rationed as follows:

- Two razor blades
- Two bars of candy
- Seven packs of cigarettes
- One cake of soap
- One package of cookies
- One pack of chewing gum per week

Had the day off but am still confined to the post. Went to buy some stationery but could get nothing except English stationery, which is tacky looking narrow tablet form stuff. Still, I had to use it, for nothing else was available.

The sky here is full of planes of all types. Found out we are in the center of a significant historical area and hope to go sightseeing before leaving. I am now in a replacement pool from which we are supposed to be sent to school as a combat unit. By the way, even stamps are rationed: ten airmail stamps or four air mail stamped envelopes per day.

*Post Exchange.

Friday, May 26th

Went through processing today and, as usual, it was a big farce: first, four big shots got up and popped off about how glad they were to have us in the Theater, then we had a short arm for a 64* and had our shot records checked. We drew our new gas masks today and checked them in the chambers. Had a six-hour pass to go into Stone tonight. That's 1½ miles from camp and we had to walk. After seeing the town, we stopped in an English pub (bar) and had a drink of cider. The pastime here is darts and, after the English explained the rules, I tried several games. The town was rather queer to me but they have some beautiful flowers. Here the movies are called flicker houses. My feet are killing me.

*One of my volunteer proofreaders explained: "Ah, the good old 'short arm!'" This was a visual inspection of the genital area by a medical person looking for any sign of venereal disease. In World War II all military personnel had to undergo this humiliating experience at various times. Thank goodness this practice has long been discarded.

Saturday, May 27th

Today was just another day, for I didn't leave the base nor have anything scheduled. Our shipping orders came out this morning and all the officers of our crew are going on one order and enlisted men on another. We shipped our bags at 1200 today; we leave tomorrow. Played volleyball for about an hour and expect to be plenty sore tomorrow. Paid one pound for my mess* since coming to this base. Found out we won't have any mail for another two weeks.

*Meals.

Sunday, May 28th

Departed from Stone by rail at 0800 and arrived at Bovingdon at 1200. These English trains run on very tight schedules. Ate my first K-rations as a lunch today; the meat is OK but the crackers taste like dog biscuits. The usual bull in signing in here. Drew site #3 for quarters and am as far as possible from classrooms and mess hall. Have to walk a mile to take a shower.

The permanent officer personnel here have their own mess and club and, as usual, they have the best of everything. Saw the Lancasters* of the RAF** go out on a mission tonight for my first time. There were about 100 heading southwest, flying very low but in no apparent formation. Very warm today, and baggage carrying is something else.

*British four-engine strategic bombers.

**Royal Air Force.

Monday, May 29th

This morning the RAF bombers went out and headed southwest apparently on a raid, for I saw about 75 — the first time I have seen this.* I am now only 26 miles NW of London but we get no passes, so I won't see the city while here. Saw General Spaatz's personal B-17 today. Several big shots fly out of here including Doolittle and Eisenhower. These paddlefeet** here are as bad as ever trying to boss us around.

Saw one formation of about 36 ships at 20,000 ft. today; these silver ships are hard to see at that height. We are to attend school here for about 14 days then draw our regular groups from which we should make our first raid in less than one week.

*The RAF flew only night raids so they must have been practicing formations or returning from a mission.

**Paddlefeet were the non-flying members of the group.

Tuesday, May 30th

Had a dental identification made today and started classes with six hours. Some paddlefoot tried to get us up at 0700 but failed in his attempt. A lieutenant finished with his missions told us we should be through within three months. Had a red air raid alert last night but I slept right on through and didn't know anything about it until this morning.

Bought a scout knife for 4 shillings then went to a softball game tonight and saw a good game. Also saw the English girls in action and it beats anything I have ever seen. They think nothing of smooching and even screwing right out in the open daylight — what a mess. I saw three couples work things over. The women look like hell and no fooling.

Wednesday, May 31st

Classes from 0800 to 2000 with an hour off for two meals and to make things worse, it was all in the same room. Now every day I see big boys going out on raids. The quota now is 30 missions but I expect it will be higher before I get through.* Saw the softball game today and, as usual, the couples were going to town all around the place. Received £32, or $129, pay. Cooler today, and it won't be dark until about 2330.

Had lots of dope on combat, ways of evading if we are shot down, how to act in a prisoner of war camp, medical aid in flight, how to bail out, and ditching procedure. From the look of things I think I have a better than average chance in the war.

*Refers to the number of missions completed before going back to the U.S. Daddy had to fly 35 missions before shipping back.

Thursday, June 1st

Another nine hours of lectures today telling us how to walk back if we get shot down in Europe. Tonight my tail is so tired of sitting down on hard chairs it actually hurts. Typical English weather today; cool and very damp with a low haze. Had a haircut, which cost only one shilling and is a fair job. Got in a craps game tonight and lost $50 as my luck went very bad, so I will try a comeback tomorrow night. The first mail to our crew came through but I didn't get anything.

Friday, June 2nd

Went to the QM* and purchased a pair of pinks**, two shirts, 12 pairs of socks, a pair of lounging slippers, two pairs of bars, two pairs of wings, and two yards of material to make my flight jacket. Saw a fair USO show tonight for good enjoyment. Got in another craps game tonight and lost $60 so quick it made my head spin. Looks like my luck is just shot and I don't think I can afford another game for a month.

Just saw three groups of B-17s returning from a raid. At noon I saw a group go over in which several ships had one engine feathered#. A P-38## made a forced landing here today when both engines cut out; overshooting, he went into the bunker at the end of the runway damaging the plane slightly.

*Quarter Master was the officer in charge of supplying quarters, clothing, food, fuel, etc. to the personnel.

**Pinks were trousers with a pinkish-gray color worn with either an olive-colored shirt, a "pink" shirt, a regular dress jacket, or an Eisenhower jacket.

#To feather a propeller is a procedure to lessen the air resistance and reduce the drag when an engine is inoperable. The dead propeller blades would be turned parallel to the wind to stop the propeller's rotation.

##The P-38 was a fighter plane. P was for pursuit, B was for bomber.

Saturday, June 3rd

As usual, school again, with everyone telling us how rough our 30 mission combat tour will be. Hit the jackpot on mail with three from my wife and one from Mama; this was my first mail. Mary Liz home OK but didn't enjoy the layover in Chicago too much. She was blue. The letters made me a little blue and lonesome for I miss Mary Liz in every known way.

Shot craps with my last £5 note and won £28 or $112 before an air raid alert broke up the game. Saw the searchlights over London combing the skies for ships. Wrote an airmail and a V-mail letter to see which is the fastest service from here home. The weather is still plenty cool.

Sunday, June 4th

Classes today as usual, and if someone hadn't told me, I never would have known it was Sunday. Saw formation after formation of our heavies* flying in large numbers. Had two letters from Mary Liz and they certainly read well. Got in a craps game tonight and lost my pants as I dropped about £25 or $100. What a foolish thing this was and now I must shoot again to get some of it back.

Had a two-hour practice air raid alert and have been informed to have our gas masks handy. See the searchlights of London searching the skies. Hear the RAF flying over and they are in large numbers. Must be something brewing.

*Heavies were bombers capable of carrying a heavy bomb load long-range — B-17s for us, Lancasters and Halifaxes for the RAF.

Monday, June 5th

Heard Rome has been evacuated by the Germans, and the Allies are now chasing them north of the city. The big birds were out again today in large numbers and some flew over with signs of battle damage. Shot craps again and broke about even; my luck seems to be shot.

Another air raid alert and extra MPs*; looks like things are about to happen. The RAF out again tonight. Colder than hell to be well along in the month of June. Four more days of school. Searchlights of London busy again tonight.

*Military Police.

2

THE INVASION IS ON

Tuesday, June 6th *

Airplanes flying over woke me about 0400 and from then until now they have not stopped. At any time you can see some of them. Saw swarms of P-38s and B-26s and large formations of 17s and 24s.** From this I thought that something was up.

About noon we heard the invasion is on and landings are being made in Northern France. Orders came out and from now on we wear our 45 pistols, gas masks, and helmets at all times. Gas wouldn't surprise me any at this stage. The landings have been made and things sound really good but there's still lots to be done. Everybody here buzzing with enthusiasm since noon. News as scarce as hen's teeth and people back home probably know more than we do. Borrowed £10 from Al tonight and lost again while shooting craps to the tune of £5.

*June 6th was D-Day.

**P-38s were fighters. B-26s, 17s, and 24s were bombers.

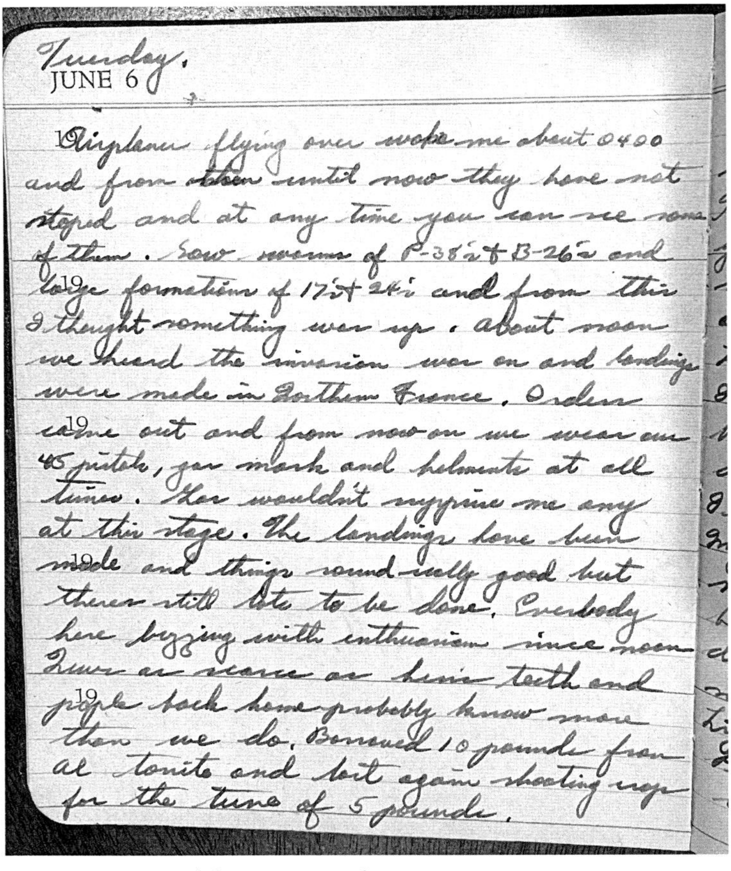
Tuesday,
JUNE 6

Airplanes flying over woke me about 0400 and from then until now they have not stoped and at any time you can see some of them. Saw swarms of P-38's & B-26's and large formations of 17's & 24's and from this I thought something was up. About noon we heard the invasion was on and landings were made in Northern France. Orders came out and from now on we wear our 45 pistols, gas mask and helmets at all times. That wouldn't surprise me any at this stage. The landings have been made and things sound really good but theres still lots to be done. Everybody here buzzing with enthusiasm since noon. News as scarce as hen's teeth and people back home probably know more than we do. Borrowed 10 pounds from Al tonite and lost again shooting crap for the tune of 5 pounds.

Daddy's journal page on D-Day.

Wednesday, June 7th
Logged 1:00 Link Trainer
(Standard Beam Approach)

Slept under five blankets last night and wasn't any too hot. My pilot, Henry Haines, received his First Lieutenant promotion yesterday, effective May 15th. Messed up two standard beam approaches in the Link* today and expect to have more scheduled tomorrow. Long practice air raid this afternoon, which I spent in an open bomb shelter.

The big boys have been going over all day in large formations and right now I hear another bunch directly overhead. My shoulder holster for my 45 pistol cuts my neck and I think I will draw a belt type as soon as possible. But it's dumb to carry this thing, for I have no rounds of ammunition. Mary Liz writes she has bought a 37 Plymouth coupe, so I now own a car.

*The Link Trainer was a flight simulator to help hone skills at instrument flying.

Thursday, June 8th
Logged 1:00 Link Trainer
(Standard Beam Approach)

Rainy and cold and just an all-around messy day. Took final exams today, and in aircraft recognition I really screwed up. But I will still see combat. Regardless of weather the heavies were out today and the invasion seems to be going in our favor but first reports show significant losses for our ground troops. Borrowed £5 from Haines and ended up a £10 winner in the craps game. Long way to go before I get even again. Had my camera seal removed and will take some pictures tomorrow.

Friday, June 9th

Finished my last four hours of class and cleared the post today. Bought some wool and cotton undershirts and a set of air corps wings. Lost £4 in the craps game. Saw the show Mission Tomorrow, which is a good show about the book by Davis. Found out Ben McLauchlin* is in the 388th group but too late to be assigned to the group, for our orders came through to the 452nd group. Some more of my tough luck.

Heard "Two Sleepy People" and it brought back sweet memories. To get our two razor blade ration per week we now have to turn in two used blades. And so is the army.

Rained all day and was cold. The invasion isn't going so hot from the news I hear. All packed and ready to move, for I understand we leave at 0600, which is plenty early.

*Ben McLauchlin was Mother's first cousin. He was a Lt. Colonel and executive officer of a B-17 airbase in England. If Daddy had known in time he might have been able to be assigned to Ben's base and could have had a more cushy job.

3

ASSIGNED TO 731ST SQUADRON

Saturday, June 10th

Shipped from Hemel Hempstead this morning and am now assigned to 731st Squadron 452nd Bomb Group of the 3rd Wing stationed at Attleborough, a small town near Norwich.* The trip up was a honey. Had dinner at Cambridge, which was about as close to an American city as I have seen in England. During my four-hour layover I saw King's College and its old cathedral, which dates back to about 1620.

Talked with three girls on the way back from Cambridge

to Attleborough and I think they were bags. Ben is stationed only about 20 miles from here, so I suppose I will see him. If I had known he was there I could have gotten in his outfit. The BOQ** here is piss poor, for we still have these three-piece mattresses and hard little old pillows.

*The name of the airfield was Royal Air Force Deopham Green. During the war the 452nd Group flew 250 missions from this airfield and lost 110 of its bombers.

**Bachelor Officers' Quarters.

Sunday, June 11th

More ground school as we had 12 hours of classes today repeating the same things we had for the past two weeks. We are allowed only two showers per week here, on Wednesday and Saturday, and to do this we have to walk better than half a mile. The group seems to have a fair combat record but it only came to the ETO* in January and they have pulled almost every long raid. Understand we are not assigned ships anymore and 30 missions are a thing of the past. Ran into Lieutenant Phillips and Lieutenant Roses who finished advanced with me and have been here since the group came over, each having better than 25 missions. Baggage lost again and no sheets on which to sleep. Had a red air raid alert last night but I slept right on as if nothing had happened.

*European Theater of Operations.

Monday, June 12th
Logged 1:00 Link Trainer
(Procedure Ascents and Descents)

Called again at 0600 and started classes at 0700 but finished with only 6 hours of classes. Watched the boys return from a raid and go through the group landing. At the same time I could see ships peeling out of formation in four other fields. The sky was full of ships, and I don't mean maybe. Tried my transformer but it doesn't work, so tomorrow I will go to the radio shop and have it checked. Baggage arrived and I sent out laundry and dry cleaning. Attended a group discussion on past missions and awards of medals being made to old combat members. The CO* estimated the cost of a trip over Europe with twelve 500 pound bombs at $27,000.

*Commanding Officer.

Tuesday, June 13th

Slept until 1000 this morning and I really enjoyed this extra sack time. Had three hours of classes. Checked transformer at the radio shop and now have my radio going just like back home. Drew a Mae West*, electric flying suit, gloves, shoes, a second oxygen mask, and had a parachute fitted. Bought my rations for two weeks and now have two ration coupons. Drew impregnated clothing for use in a gas attack. Expect we will make our practice mission tomorrow and go on our first mission the following day. Cleaned my pistol and now have it ready to go with 20 rounds of ammunition. Rained all day.

*A life jacket was called a Mae West as it made the men look like Mae West with a full buxom figure. Later they were referred to as a Dolly Parton.

Wednesday, June 14th
Logged 1:30 Practice Formation (Low-altitude)

Made my first flight over this section of England and I have never seen so many ships or airfields. After landing our hydraulic system went out and, after nicking a wing on one ship, I managed to pump enough pressure on the emergency hand system to stop before we plowed into the tail of another ship. Shot skeet and I only broke 17.

The 452nd put up 25 ships today and every ship had a least one flak* hole when they returned after making three passes at the target. A Halifax** attempting a forced landing here crashed and burned off the southeast corner of the field. Only two came out alive and they aren't expected to survive.

Took a warm shower and did it ever feel good. Had a letter from Daddy and, as usual, it made me rather homesick. Went to the show.

*Flak is antiaircraft fire.

**The Halifax was a British four-engine heavy bomber used throughout the war.

Thursday, June 15th
Logged 2:00 Practice Formation (Low-altitude)

Made roll call at 0830 but didn't have anything, so our crew beat Martin's in a game of softball; that ball certainly hurt my hand. Had an easy mission scheduled for my first one today but it was scratched. Flew practice missions out around the shores of the North Sea and never have I seen so many airfields. Saw the barrage balloons at Norwich, which were pulled down. Called Ben McLauchlin tonight. The boys flew an easy mission today.

Friday, June 16th
Logged 0:30 Link

Was supposed to fly a couple of hours of Link but a change of schedule cut me short. Stopped by and picked up my escape and evasion photos and they are lousy. Had a medical lecture on venereal disease and delousing; as usual, the doctor painted a horrible picture and recommended that the boys keep their tools in their pants. Received two letters from my wife written January 17th and 21st at Grand Island, through the U.S. Army mail service. Saw a C-47* give the control tower here a real hot buzz job. No missions were flown from this base today.

*C-47 was a military transport aircraft.

Saturday, June 17th
Logged 3:30 Practice Formation
(10,000 Above Overcast)

They pulled me out of bed at 0030 to fly my first mission after only about two hours of sleep. Had fresh eggs for breakfast and went to briefing. Hurried out to the ship but forgot my Mae West and parachute harness and had to come back for them. Loaded with twelve 500 pound GP* bombs for an airfield about 20 miles northwest of Paris. But at 0430, just before takeoff time, the mission was scrubbed. Went back to the sack.

Flew a practice formation of 14 ships above overcast, making an instrument assembly and a peel-off with instrument descent, flying through an overcast. With so many ships in the area it really makes you sweat. Had a good hot shower and shave for my regular Saturday bath. The possibility of my first mission didn't bother me and our crew seemed relatively calm as if going on a routine flight.

*General Purpose.

Sunday, June 18th
Logged 7:30 High-altitude Formation (27,000 ft.)
Hanover, Germany

Called at 2400 for 0100 briefing and they weren't shooting blanks. Briefed to bomb oil refinery about four miles west of Hanover going out over the North Sea and south to the target. Crew very calm and all set by takeoff time, even I had everything. Flew ship 780, "Windy Lou" but had to wait on mechanics to fix #1 engine* before takeoff.

On takeoff the flair gun fell from the opening and shot a flair in the cockpit; Hank throttled back as both cockpit windows opened, then poured on the coal to take off anyway as Lundstrum put the fire out. The fire burned out the hydraulic pump. Took our position as #2 ship of the high element of the high group of the formation. The entire trip was over an overcast, and we dropped on the PFF** ship.

Saw more flak than the law allows on our bombing run but it was mostly off our right wing. Only saw American fighters. Could hear small pieces of flak bouncing off the ship. Trip back not eventful, crew very quiet, B-17s were in large formations ahead as far as the eye could see. Had to circle traffic pattern three times while Perry fixed the ball for landing; this gave me lots of time to build up pressure with hand-pump. Slept all afternoon and went to bed about 1900 and in nothing flat I was sound asleep.

*B-17s had four engines. They were numbered one

through four when looking forward.

**Pathfinder Force was a primitive radar system. It was a B-17 especially equipped with radar navigation for targeting through heavy cloud cover, so the other planes dropped their bombs when this plane did.

Monday, June 19th
Logged 5:30 High-altitude Formation
Feathered One Engine, Aborted

Called at 2400 for 0100 briefing. Target was an airfield near Bordeaux, France; took off at 0510 and made instrument assembly at 9000 ft. flying 360 "Smoky Liz." We were in formation until we were ten miles inside France at which time our #2 engine went dead and we were forced to abort.

Sergeants Lundstrum and Claybaugh failed to show up, so we took off with only eight men. Had trouble staying in formation pulling extra power all the way; had trouble with #1 and #3 when #2 went out, bango. What a crate. Landed with our bomb load of 20x250 pound GP. The mission was scrubbed after we aborted. Solid overcast.

Called at 1530 after two hours of sleep for a briefing but this mission was scrubbed before we were briefed. Critique tonight but our mission of yesterday wasn't discussed because we flew with the 96. So far, we have only been assigned beat-up old crates, which should be discarded.

Tuesday, June 20th
Logged 8:30 Alt. Formation (27,000 ft.)
Lost #2 on Bomb Run
Magdeburg, Germany

Called at 2400 for 0100 briefing and drew 360 "Smokey Liz" again. Target a synthetic rubber plant at Magdeburg, Germany. Bitched like hell and swore we wouldn't get back in the locker room. At 0100 the weather was clear but at 0415 there was an overcast solid when we took off. Saw Germany from 27,000 as they laid down smoke screens over numerous targets on our route into the target. Number 3 running rough and hot until deep in Germany. Making the bomb run with an extra boost to stay in formation, our #2 went out and we fell out of formation right before bombs away. Flak over the target was intense and we did evasive action but flak was everywhere. Lucky, we only picked up two small pieces.

Lost our formation, tacked on to another group. Couldn't feather, for the oil line to feathering mechanism was out. Put prop in full low RPM and left her windmilling* at 1100 with the gas cutoff switch closed, switch off. Tagged on to three different formations before finally making the North Sea and followed behind the fourth to come on home. Really got pissed off when we flew right along the coast coming back; we were meat for fighters.

Thought the boys would have my things divided when I got back, but they didn't.** Flak at the target was aided by some rocket shells, which came up, failed to explode, and went back down; these were visual. Found out we got credit for yesterday's mission, so I now have three. Slept all afternoon. Colder than hell tonight. Found out tonight we are scheduled to go to Russia on a shuttle mission.

*Windmilling is a procedure to lessen the air resistance and reduce the drag when an engine becomes inoperable.

**Sometimes, when an airman was shot down or bailed out, the other airmen would help themselves to the missing person's belongings.

4

ALL HELL BREAKS LOOSE

Wednesday, June 21st
Logged 12:00
(Alt. Formation 21,000 ft.)
Ruhland, Germany

Called at 2330 for 0130 briefing and was told to bring Class A* uniforms and enough other clothing to last a week, for we were headed for Russia. Bombed a synthetic rubber plant at Ruhland, Germany with 42x100 pound incendiary bombs, and the results looked good, for when we turned at the rally point** smoke was pouring up from the target. Flak over the target was moderate.

Saw my first dogfight when a P-51# took after an ME109##, which made a pass at our formation and, after a short chase, shot the German down in flames. After this fighter attack, the formation was flown much tighter. Landed on a metal landing strip at Poltava, Russia, at 1700 after pumping gas from one tank to another. When we landed, there were only about ten gallons per engine left. Our P-51 escort flew all the way with us but we brought three external tanks in each of our ships, so they can use them going back. We picked up only three flak holes. For the first time, 8th Air Force bombers shuttled from England to Russia. Flew over the old battlefields and the trenches, roads, etc., could all be clearly seen from the air, and points of fierce fighting stood out like a sore thumb. The towns over which we flew showed the results of two great armies retreating through them, for they were all burned out.

When making our traffic pattern over Poltava, I couldn't see a single building that hadn't been gutted by fire or bombs, leaving only the brick walls standing. The Russians are amiable people. They were astonished when our bombardier, Lt. Bebenin, started talking with them in Russian, and from the sound of things, they were having quite a good chat. Before we left the ship, a German reconnaissance plane flew over, and the Russian antiaircraft fire went into action but they didn't get him. Was awfully hungry but couldn't eat much of the food, for it was really a piss poor meal. Had some cookies that were too hard for me to eat, some English stuff.

Assigned to Quarters, which was a tent with eight men in it sleeping on army cots with no pillows and only the ground for a floor. We have slit trenches for latrine, and there are no showers or hot water; to shave, they have basins and a limited supply of cold water, which you draw from a water trailer. Turned in early, tired and sleepy, cutting a lecture we must hear before going on pass.

*An "A" was the class A uniform of a shirt, trousers, blouse, and tie.

**The rally point was an area out of range of enemy flak batteries where the squadrons could reform into combat wing formations.

#P-51, our fighter.

##Messerschmitt ME 109, the backbone of the German fighter force.

Thursday, June 22nd

At 0030 this morning, all hell breaks loose around here as the Germans turned loose flares to light up the field, the Russians threw up flak at them, and I left my tent running like hell for the slit trench*. Didn't know exactly where they were but knew they were in a large field, so I ran like hell across this until I fell in one where I stayed. For the next two hours, I saw one of the most concentrated air attacks possible as the Germans went after our B-17s. Most of the bombs hit on the airfield, and there were numerous explosions; and then the ships started going up in flames, the gas dump exploded, and two or three bomb dumps went off. All this time, I was sitting up like a damn fool watching the fireworks, and for this, it was a good show as antiaircraft shells exploded, rockets and tracers filled the sky, flares floated down on parachutes, and bombs flashed. Some bombs hit within 100 ft. of me, but when a 2,000 pound one went off about 400 ft. away, the earth really shook. As I heard the plane very close in both cases, I was well down in my fox hole. Several men were wounded, and only luck prevented me from the same because I watched the entire time without even a helmet, and the metal was falling fast and furious all around.

After a lull, I started back to the tent but Jerry** changed my mind; as he returned, so did I. Finally went back to the tent and after watching the ships burn and shooting the bull went to bed and slept till 1000. Wanted to look the ship over but we were restricted from the field, for both delayed action and antipersonnel bombs had been used, along with almost every other known type of bomb. From the boundary, I could see better than half the ships had burned to the ground. Out on the field, Russian women were busy locating unexploded bombs and detonating them or putting the smaller ones in a pile to be detonated together. What nerve and guts. From reports, the antiaircraft fire brought down seven Germans. The Russians captured a German airman who had jumped and promptly escorted him to Headquarters at the point of a bayonet.

All day long, there was the boom of the exploding bombs. After sweating out a long line at supper, just before being served, a red alert sounded, and along with everyone else, I took off to the woods. But only a reconnaissance ship flew over, and antiaircraft fire opened up. Evacuated camp, and the entire outfit is sleeping in the woods under the clear starry sky. Lieutenants Bebenin, McCann, and I took a walk and ended up in a cherry orchard where we ate our fill of good big ripe cherries. Here the government owns everything, and every man and woman put in six days of work per week for the government. From what I saw last night, I will stay on the sending end, for I never again want to be on the receiving end of a bombing. Last night we had a pilot killed outright and a copilot who isn't expected to live. The Russian soldiers took a terrible beating when they tried to put the fires out, and I hear about fifty were killed.#

*Besides a latrine, it is also a narrow trench for protection like a foxhole.

**Slang for German.

#For more information on this raid, please turn to the Epilogue, page 177. (I later learned that Daddy and his crew were one of 26 crews from the 452nd sent on this mission.)

Friday, June 23rd

Caught in a light rain before getting up but only put my clothes under the cot, pulled the blanket over my head, and went back to sleep. Checked the ships, and I must say the Germans did a damn good job on this field. There are 32 B-17s burned right to the ground and another 19 so severely damaged they will never fly again and, besides these, they got two C-47s and one P-38, all because some big shot had his head up and locked. Our ship didn't burn but it has around 200 holes in the fuselage mainly, all three tires are flat, two engines have slugs through the pistons, one oil tank was punctured, and she will require about two months to repair before flying again. It was an awful sight to see four burned engines, the ends of the wings and the tail assembly from the entrance back, and a pile of ashes and melted metal between these to represent what was a B-17 only a few hours before. I salvaged myself a good flak helmet and picked up several souvenirs. The totals show about 40 Americans wounded by flying metal.

Went to Poltava and saw a town gutted of everything where people were busy rebuilding. These people live under anything that gives them protection from the weather. The Russians are a very religious group, and there are many churches in this town, most of which are in ruins. But the first thing they repair is the churches. Almost no Russians can understand English, and as Lt. Bebenin spoke Russian with them, they all stood around blocking the street listening in surprise at an American who could speak Russian. Every time he talked with someone, they always asked how he came to know their language. His parents are Russian. From what I could gather, the town looked relatively modern with nice buildings, which were wrecked when the Germans withdrew. One Russian captain told us the Germans gave many local girls some form of venereal disease before leaving to slow down the Russian drive.

Returning to the woods late this afternoon for another night outdoors. Went swimming in a small lake and washed off well for the first time in about five days. While busy washing with soap, four Russian girls came by. When we motioned for them to come in, much to our surprise, off came their clothes, and into the water, they came, in the nude. Couldn't say anything to them but they caused hard feelings, and I have never seen such big firm tits. Couldn't do anything but look; loaned them our soap, and after a good bath, they left and went on about their business as if nothing had happened.

Went to a folk dance at a crossroad school tonight and saw the Russians dance to a tune from an accordion. Enjoyed seeing this.

Saturday, June 24th

Almost froze last night, and when I woke up this morning, I was covered with dew. No air raid last night. Moved 65 miles from Poltava west to Mirgorod by LI* truck over the roughest country road I have ever ridden over. My guts are sore from this trip. This was the main road but it wasn't paved and, due to a labor shortage, they haven't worked on it much in the past three years. All along the road, there are signs of war as you see wrecked tanks, field guns, blown-out bridges, and large numbers of graves; Russians are buried where they fall. The fields are all green with grain crops, and the Russians expect a good harvest this year.

The farmers are busy rebuilding their homes, which are made of mud over a frame of tree branches with a straw roof. Women and older men are doing all this work. These people make their bricks by baking mud in a molded frame in the sun. Women here are doing most of the work on the home front from plowing with a team of oxen, repairing the roads with a shovel or drag behind oxen, even rebuilding blown out bridges on the main roads, or any other work needed. Saw men sawing lumber by hand as large logs are pulled up on a sawing stand. With one man above and one below to pull the cross-cut saw, they do an excellent job of sawing planks.

We are now quartered in an old Russian brick barracks, which is fixed up nice but we are still eating out of our mess gear. The town here was not totally destroyed, and there is a big Russian hospital here. I see lots of their patients about town. We were moved over here to help load bombs in the B-17s for the bombing shuttle run back to Italy. Still don't know how we will return. Went to town tonight and found there are many Polish people here. Visited the Greek Orthodox church, and it was interesting, for the church was all repaired, and the interior was painted with different bible scenes. When you talk with the Russians, they always tell you where the women can be found, and they all seem eager but I am laying off, for they don't exactly appeal to me. They are all well built and have nice-looking faces but they are not very clean due to a shortage of soap and water. In peacetime, I understand they are very clean people.

*Light truck, probably a one-and-a-half-ton. The U.S. provided thousands of them to Russia during the war.

Sunday, June 25th

A sack time day with nothing to do but sleep and have a good sack with a mattress and, for the first time in three nights, sleep under a roof. Wrote letters and played a little softball. The Russians eat roasted sunflower seeds the same as we eat peanuts back home. The Russians are a very religious people, and on Sunday, no one works; but they all go to church without fail. In the afternoon, they had the marketplace going full blast, with all trades being made by barter. Took a short walk by the river and walked upon two women taking a sunbath in the nude. Spoke, but didn't stop, taking a damn good look as I passed. Saw German observation planes overhead four times today, so everyone expects an air raid tonight. The B-17s are here now but are flown out just before dark each night.

Monday, June 26th

No air raid last night, so I had another good night's sleep. The enlisted men loaded the bombs in the Forts*, and at noon today, they took off for Italy on another leg of the shuttle run. Watched two men sawing lumber by hand this morning. At 1300 we loaded into trucks for the rough ride from Mirgorod back to Poltava, which was a dusty five-hour ride and plenty rough. Many GI truck drivers could take some pointers from the Russian drivers, especially the jerk who brought us over here today. They have started cleaning the wrecked ships off the field, and the metal landing strip has been repaired. There are several C-46s and C-47s here, so it looks like we go back by ATC** and the long trip around the Mediterranean Sea. Tonight they are still exploding bombs dropped by the Germans in their big raid here the night of June 21st. The trip was a return on the same route and noted nothing new.

Went to Poltava again tonight. Money here is worthless as we learned ice cream sells for $5 and $10 per cone, a loaf of bread is $10, and a package of American cigarettes is worth at least $15. The base was putting toilet paper in the latrines but it kept disappearing, and a close check showed the Russians were taking it to roll cigarettes. They are short of paper, and after taking a crap, they merely pull up their pants and go on about their business. Saw a locomotive much on the lines of the English trains with a woman as an engineer. These women here take on all jobs.

*Forts were B-17s — Flying Fortresses.

**Air Transport Command. They hauled cargo and passengers, not bombs.

Tuesday, June 27th
Flew 2:00 as Passenger in C-46

Sacked most of the morning, then tried to borrow a 35MM camera with no luck. At 1400 we were called together and told to pack. At 1500 we were loaded aboard a C-46*, and after circling Poltava, we headed southeast. As we departed, they were still setting off dud bombs.

Flew to Rostov, where we were called in and are staying for tonight. Bunked down in a Russian barracks sleeping two to a single bed and ate my first real Russian meal of meat, fried potatoes, beans, black tea, rice cooked with dates and other spices. For this meal we were called out of bed at 2330 after thinking we weren't going to be fed and turning in. The Russians were about to get pissed off when we were slow about getting up. We left Bebenin at Poltava, where he stayed as an interpreter. The battle shows on this town.

*C-46 is a military transport plane.

Wednesday, June 28th
Flew 1:40 as Passenger in C-46

Left Rostov at 1100 and flew down to Armavir. This is a training base, and they shoot landings all day, never leaving the traffic pattern, and are these new boys making rough landings. There are some good-looking Russian women here. We had a big Russian dinner tonight with all the trimmings, which started with a water glass full of vodka as an appetizer. Allen really got high on his drink. Everyone felt no pain by the end of the meal, but some made a hog of themselves and got plastered; the Russians looked at them with a frowning eye. Walked uptown and everyone welcomed us by giving us bouquets of flowers, cherries, apricots, etc. Had a big dance and a Russian show tonight.

Thursday, June 29th
Flew 5:30 as Passenger in C-46

Takeoff was delayed, so we walked uptown again, and everyone offered us cherries, apricots, and flowers — such a heart these Russian people have. Finally given an instrument clearance and took off just after noon, flew over an overcast until after we cleared the mountains, then contact across to descent until we reached Tehran.

Coming in, I noticed across the desert straight rows of holes that resembled doodlebug holes, only much larger, which I found out tonight was an irrigation system many years old. These holes are dug by hand, and below the surface are connected by a tunnel through which the water flows, thus reducing evaporation from the sun's heat. Some of these systems are about 1000 years old. From the air, Tehran looked rather nice as each building had its roof of many domed sections.

The town is a mess with its mixture of every race of people and dirty-looking shops. Here the water system is a stream of water flowing between the sidewalk and the street; in this, I saw a woman washing clothes, a man taking a piss, and within less than 20 ft. downstream, two women filling pitchers, which looked like their supply of drinking water. The lower class people here sleep by this stream, for it's cool outside after dark. The higher class have high rock walls around their places so no one can see what goes on inside, but some of these places were really nice from what I could see. A U.S. officer told me some of these women are beautiful, but a girl is automatically considered a prostitute when seen with any American.

You can buy anything here, for this is a neutral country, but prices for a U.S. uniform are plenty high. I bought a couple of souvenirs. The serfs riding on their asses look very funny, for their feet almost drag the ground. I see asses used as pack animals and driven in trains. The Officers' Club at this ATC base closed early tonight so they wouldn't have to sell us transients their whiskey.

Friday, June 30th
Flew 7:15 as Passenger in C-54*

Called at 0530 and took off at 0815 with some TO** pilot who bitched about everything: the weight, the center of gravity, etc. Rode through two takeoffs and landings, which were all piss poor — no wonder this pilot is with the ATC. The trip was interesting over the desert, and I enjoyed seeing the isolated towns with their high city walls, miles from rail service or another town, with only winding dirt roads leading into them. The thought of how they got supplies amused me. Could smell gas fumes, which worried the pilot, so we landed in Iraq just west of a river to have a look, which took about two hours. The place felt like it was on fire as the temperature reached a hot 125 degrees Fahrenheit. Stepping out of the ship was like stepping into an oven when the hot air hit you in the face.

Flew real low over the Dead Sea, River Jordan, Jerusalem, and Bethlehem and got a good look at these places of Biblical interest. This is more of an oasis in the desert. The Dead Sea has a greenish-blue color, and salt evaporation pits on the north shore stood out very clearly. Saw the Mediterranean Sea, Suez Canal, and the Red Sea.

The Suez Canal looked like a big drainage ditch right through the desert, for there was nothing but sand on each side, and I didn't see a lock on the entire thing. Made a large circle around Cairo and got a good aerial view of the pyramids and sphinx and the city of Cairo and the surrounding Nile River valley. The pyramids are enormous things that stand out like a sore thumb from the air. Saw a B-29 on the ATC field here. This Cairo base is really OK, for after a good hot shower, my first since leaving England, I ate an excellent hot meal served by negro waiters in a nice club. Confined to the base and couldn't buy mixed drinks, for we are transient officers, but enjoyed drinking six or seven cold Coca-Colas. Bought some souvenirs at the PX.

*C-54 was a military transport plane.

**Transport Officer.

Saturday, July 1st
Flew 12:45 in a C-54 as Passenger

Called out just before sleeping to leave, and while waiting, I talked with the guard who gave me the low down on Cairo. Departed at 0045, and shortly after takeoff, everyone was asleep on comforts furnished by the ATC. Woke before landing for gas and oil at Tripoli about 0645. The signs of battle were everywhere. Trenches and discarded equipment could be seen from the air, junked planes and riddled hangars from the ground. Departed 0800, saw the Mediterranean for a stretch with orchards along the shore then more desert wasteland with nothing for miles. Landed in Casablanca at 1445, and is this a large base.

Rode the bus into town and back, but didn't stop. The town looks very much like Florida. But there are lots of strange-looking people. The French people here are beautiful. At restaurants, food and beer are served right out on the sidewalk to tables placed there. The buildings are modern and look clean from the outside. Would like to stay here overnight. Had to go to the toilet, and was that French job ever a surprise, just like a good old country job, only here there was a square contraption laid level with the floor with footrests on the sides. By placing your feet on these and assuming a full knee bend position, one went on with their work. Saw several battleships lying at anchor. Waited three hours for the ATC to get in a mood to go.

We are now heading for England, and I am just about ready to stretch out for a good sleep. These C-54s are good ships, cruise at 170mph, have lots of room, and the crew has bunks to sleep in on these long trips. All the enlisted men and I are not together; left Lt. Haines in Casablanca, for he was too far down the list. Lt. McCann stayed in Cairo, for he was rather sick and didn't feel like flying.

Sunday, July 2nd
Flew 9:30 in C-54 and 4:00 in C-47 as Passenger

Landed at Prestwick, Scotland, early and had breakfast in an old castle, which is being used by ATC as an officers' club. Really makes a nice one. Took a good hot shower, which felt good. Walked around the lawns and forest of the old estate for a quiet and enjoyable morning. See by the New York Times where the German raid on Poltava was light with minor damage; what a crock. I would hate to be around during a heavy raid.

Flew to Harrington in a C-47 through the soup and came the other 30 miles by GI truck. Coming over England I saw these fields covered with gliders and C-47 tow planes, so it looks like another beachhead. Had 20 letters here when I arrived, so I spent the night reading. Told the boys in the hut some of my experiences and gave them the right dope on the bombing.

Monday, July 3rd

Raining and messy all day. The group has pulled only two missions since we left. Lt. Anderson and his crew who came over with us landed in Sweden the day we went to Russia. I got paid better than £34. Talked with G2* and we are not supposed to say anything about the raid. Returned Martin's blouse. My first laundry came back five pairs of socks short. Turned in the escape kits two short and the binoculars short, and the G2 officer just wrote the stuff off. Tried to arrange the enlisted men's pay, but they must wait until the 10th for supplementary. Looks like we may have a pass coming up in three or four days. Haines came on in today. It's good to be eating out of dishes again and have something prepared well, like the fresh eggs we had for breakfast. After so much powdered and canned stuff, this is very good.

*G2 was military intelligence.

Tuesday, July 4th

A relatively quiet day for the fourth with nothing happening. Had ground school all morning on engine operation. Drew some new flying equipment and had a mic put in my oxygen mask. Today was gas mask day and they took my name for no mask, so I told them it was in Russia. Put some of my collection of paper money in a cellophane protective case. The group pulled a raid on a bridge below Paris, but after four runs had some flak and returned with the bombs. We won't be operational until McCann returns. Wrote my wife about the Russian women in the nude.

(Daddy said in an interview many years later that July 4th was a day of thanksgiving for them, to be alive and at their home base.)

Wednesday, July 5th

Ground school on engine operation. Spent the afternoon fixing my paper money collection into Plexiglass containers to send it to Mary Liz and explain what the stuff is and its country. Received my pin representing a division citation given to the 3rd Division of the 8th Air Force. Wrote several letters then turned in for the night.

Thursday, July 6th
Logged 6:00 High-altitude Formation (25,000)
Abbeville, France

Ground school all morning on engine operation. Alerted for a 1415 briefing; target, fuel dump for rocket bombs only a short distance inland from the French coast, 12 miles northeast of Abbeville. Drew "Smokey Liz" 360 again and all afternoon we were never in formation. Saw quite a bit of flak but it was inaccurate. The bombs were mostly out of the target area, but some fires were started. Landed at 2215 and I am plenty tired.

Flew a 14 ship formation and we were #4 high squadron. Flew without our regular bombardier and navigator. Saw London for my first time and though it's a large city, there are lots of parks and things don't look too crowded. Millions of barrage balloons south of London protect the invasion supplies from low flying planes.

Friday, July 7th
Logged 9:00 High-altitude Formation (25,000)
Leipzig, Germany

Called at 0045 to fly another mission and did I ever feel bad; and besides, I didn't have time for breakfast. Target was some synthetic oil plants about 15 miles SW of Leipzig, Germany. Bombing results looked good and the flak was heavy and accurate. Our ship picked up 12 holes. A piece barely missed our radioman Sgt. Ackerman as he stooped for some more chaff*. Flew #4 position in the high squadron low group. Tired and pissed off on this mission. Slept all afternoon. Critique tonight and all the fellows tried passing the buck again. Hope to hell I get a rest tomorrow. This makes six missions and I am now eligible for the air medal.

*Strips of foil dropped by an aircraft to confuse enemy radar by creating false blips.

Saturday, July 8th

Caught up with my sleep this morning by sleeping until 0900. The mission today was supposed to be a milk run* until they got the hell shot out of them. The group lost three crews. Smith's crew from our squadron went down on his 4th mission.

Shot skeet and I pulled a very low score of only ten, for I was really off. Tomorrow our crew goes on its first pass since arriving in England. Started over to see Ben McLauchlin tonight but couldn't get out of Attleborough. This day off came at the right time for I am rested at last.

*Milk run was an easy mission.

5

GOT THE HELL SHOT OUT OF OUR SHIP

Sunday, July 9th

Got up in time for breakfast, shaved, and packed for my two-day pass. All the crew got enough money for a pass and are off. Talked with a 1st Lt. who knew Harold Hatchett* and came over with him from Camp Gordon. Stopped by and saw Ben for a short talk. He seems well set and enjoying life as an S-2 Lt. Colonel.

Staying in Cambridge and have a nice room with a good soft bed and sheets for a change, so it's a nice long sleep tonight. Ate supper at the Red Cross Officer's

Club. Took a walk about town looking over the old buildings and odd sights. This is a very old town with some buildings dating back to about 1000 AD. Ran into an old buddy from Sioux City, Lt. Douglas, a navigator from the pool.

*Harold Hatchett was Daddy's best friend since childhood. The families lived across the street from each other. He was injured on Anzio Beach, Italy, as our troops were invading. A hand grenade was thrown at his group. He covered it with his helmet and threw himself on top of the helmet. He died soon after in a hospital nearby and is buried in Italy.

Monday, July 10th

After a good quiet night's sleep with a feather pillow, I caught the train to London and made a quick tour of some of the town. Went to the QM and looked over the store's supplies. Ran into Lt. Wonderling and crew; they now have quite a few missions. Looked around town for an anniversary gift for Mary Liz, but couldn't find anything and ended up with only a card. The prices are plenty steep when an American officer steps up.

Heard two buzz bombs* go over. When they exploded, I only heard a dull thud because they were so far away. Returned to Cambridge for the night. Next trip I am going to London for my entire two-day pass regardless of buzz bombs. The city looks very much like our large American town as life goes.

*Early cruise missiles.

Tuesday, July 11th

That sleep last night was wonderful. Went uptown and looked around for a couple of hours. Saw an English cucumber, which grows on trellis' over here; it was about 18 or 20 inches long. Am going to send some of these seeds to Daddy. Caught the train to Norwich. Made arrangements with Stone, who is a tailor, to have a blouse* made. Ran into Lt. Winnette from back at Ft. Sumner, New Mexico, and he has completed his combat tour in B-24s. Tried to get a leather camera case but had no luck. Visited the Norwich museum, which was very interesting and I must make another trip when I have more time. Went to a leg show at the Hippodrome Theater, which was fair, but this English humor kills me. Ran into a couple of girls and after a little talk, I caught the train back here at 0900, getting in here rather late.

*A blouse was a jacket — waist-length like an Eisenhower jacket.

Wednesday, July 12th
Logged 10:00 High-altitude Formation (24,000)
Munich, Germany

Called at 0400 for briefing and found out we were going to Munich. We were given instructions on how to go about landing in Switzerland, a neutral country over which we would fly. Flew #2 in the high squadron of lead group. The flak over the target was plenty heavy but inaccurate; we picked up several small holes. Lots of boys went to Switzerland. One B-24 pilot asked for fighter escort to the Swiss border, saying he was low on gas with 900 gallons. He didn't get fighter support but went on anyhow. We were on oxygen for about 7 hours, landing here at 1800. One ship crash-landed here after running out of gas. Bomb load 4x500 pound GP and 6x500 pound clusters of incendiaries. Enemy fighters hit the wing right behind us on the return trip. Official reports say we lost 25 heavies in today's 1200 ship raid on Munich.

Thursday, July 13th
Logged 9:15 High-altitude Formation (25,000)
Munich, Germany

Briefing at 0130 and we were a spare, but as we

crossed mid-channel we were called to fill in #6 of the low squadron of the lead group. The wing lead today by Major Roland was piss poor and staying in formation was damn hard. Bombed in wing formation and the flak today was light compared to yesterday. Had a strange new bombardier and navigator who were both flak-happy* and got the enlisted men all upset. Finally had to tell them to shut up. The Alps were beautiful, sticking up through the clouds with their snow-capped peaks and timber lines clearly visible. Could see Switzerland clearly and it was quite a peaceful looking place. These long missions with long briefings are cutting down on my sleep and am I ever tired tonight.

*When someone is mentally affected by flak, thereby making the rest of the crew nervous; for instance, by talking too much or being negative or freezing up on the job — not operating at peak performance. Some had emotional breakdowns and some were psychiatric casualties.

Friday, July 14th
Logged 9:30 Formation (18,000 and Below)
Aurillac, France

Called for a briefing at 0030 and went to a field down in the mountains not far from Aurillac where we descended to 500 ft. and dropped supplies to the French underground in containers let down by parachutes. The parachutes opened in the bomb bay but on the bombing run Ackerman, Lundstrum and I fixed them so all but one opened OK. The chute on this one wound around the bomb bay drive shaft, so I cut the lines and it fell free. Next time I will know how to handle a load of this kind. It felt funny walking around the catwalk with nothing below but the ground some 500 ft. away. The bombardier flipped the switch and the right door tore off the hinges but, fortunately, it didn't come loose. It just hung open on the return trip.

This section of southern France looked beautiful with its numerous low mountains, small lakes, and streams, which show no signs of the war. Before we landed, the control tower gave us a red flare and around we went, but we called the tower and told them our trouble and came right on in.

This trip today was something of a morale builder for the French as July 14th is their independence day or some such big occasion. This is my first mission on which I saw no flak or enemy fighters. Outside of it being a long haul, it was a milk run from the start. Lt. Hatrum finished his missions and gave the field a good buzz job, and now they are trying to court-martial him. This dirty bunch of rank here should be kicked out of the Army on their asses, though some of the boys with rank are really OK. These others couldn't make a living in civilian life. The load was 12 containers, which weighed about 300 pounds each.

Saturday, July 15th
Logged 3:20 Practice Formation (10,000)

They didn't call us this morning, for a change, and it certainly felt good to sleep until noon. Took a good hot shower, my first in a week, which made me feel much better. Started writing letters to catch up with my correspondence, but this was halted at an early stage when Operations called.

Flew copilot for Lt. Greeswell, whose copilot is in the hospital. I flew most of the time as we were #3 ship of the lead element, an easy spot to fly. The flight carried us past supper and they didn't hurt themselves feeding us. From the look of things a flying officer is about as low as anything can be, for we take it from everyone. Someday I will get fed up with all this and really tell some of these big-shot playboys a few things free.

Sunday, July 16th
Logged 3:20 Practice Formation (8,000)

Thought for sure I would go on the mission today, but was surprised when I woke up and found out it was 0800. Called at 0900 to fly a practice mission and we flew #3 of the lead element, an easy position. For this week we are non-operational and will attend ground school. Am really pissed off tonight, for our passes have been canceled because we failed to pay our Officer's Club dues. I didn't even know about the thing. Besides, the club here is for permanent officers and we combat men seldom have time to go there. Besides, we have our separate mess, which is just that, a hell of a mess. We have a party tonight but I don't feel like attending, for I am rather tired.

Monday, July 17th
Logged 4:00 Practice Formation (10,000)

This week we only fly practice missions and attend ground school. Reported to a 0900 briefing for a practice mission and we drew #5 position of the low squadron. With all of the greenhorns* this was one hell of a position to fly. We made a big circle over England after making an instrument assembly. Coming in, the captain in charge ordered us to peel off over a B-24 base about ten miles from here by mistaken identity. There are thousands of bases in this section of East Anglia. Slept all afternoon and having written several letters, I am turning in for the night. For a change today, they issued us a good one-piece mattress. I put on the two sheets, which I bought in London last week. Now all I need is a pillow and pillow slip. These sheets sure feel good.

*Inexperienced person.

Tuesday, July 18th
Logged 6:30 High-altitude Formation (23,000)
Kiel, Germany

Called at 0100 for briefing and were sent after the E-boatyards* at Kiel, but due to an overcast and an inoperative PFF, we could not drop the bombs, so we brought our 10x500 pound GPs home. Saw some inaccurate flak and a couple of rockets. We had plenty of fighter support today. This was OK until we returned and had an 18 ship formation below a 1500 ft. overcast and, with the sky full of ships, this was somewhat nerve-wracking. What a way I picked to celebrate my 29th birthday and I hope I spend all my others at home in the States. Tonight we had another meeting at which our past missions were discussed and the big shots passed the buck. This puts ten missions behind me and I hope to hurry up and get the others.

*E-boats were German torpedo boats, which were suited for the open sea with a range of 700 nautical miles.

Wednesday, July 19th
Logged 8:00 High-altitude Formation (24,000)
Schweinfurt, Germany

Called at 0100 for briefing and we were sent after ball-bearing plants in the city of Schweinfurt. Flew copilot for Lt. Lloyd, who finished his tour of duty today. Lt. McCann and Sgt. Ackerman of our crew also flew on this makeshift crew. I flew most of the formation, for Lloyd hadn't had any sleep, so he slept most of the mission. The bomb bay doors failed to open and had to be cranked down by hand, so we were late getting our bombs away. Lt. Jamerson, the bombardier who has flown on my last four missions, was just as flak-happy as ever, and over the target I had to tell him off properly to shut him up. Hope he never flies with our crew again. Flak over the target was moderate, but pretty well on our group, though no one went down; our ship had six holes. We started out flying #5 high squad low group, moved to #5 of lead squad low group, and later moved on up to #4 of the same bunch. That #4 is an easy position and I wouldn't mind flying there all the time. While we were going to the briefing room the Germans were flying around this section of England, and they dropped some fairly close, for we could see the flak and feel the earth shake. The searchlights picked them up but it appeared no antiaircraft guns fired on them. Took a good hot shower, ate supper, and am now in bed and about half asleep.

Thursday, July 20th

Got a good night's sleep for a change before attending ground school all day. Went to Operations and asked if we could be assigned a regular bombardier other than Lt. Jamerson and was told we could. Sent my honey a $15.00 money order in an anniversary card and asked her to buy a nice leather pocketbook and I will send her an 8th Air Force patch, a pair of wings, and a bar to put on it. This ground school is certainly a crock, and to think I missed an easy mission today on which the boys didn't even see flak. Collected some English coins for Elizabeth* and fixed them for mailing. Am trying to fix a box for my honey. Bought my weekly rations.

*Elizabeth was Daddy's sister. She was pregnant with her first child when her husband was killed in the war.

Friday, July 21st
Logged 9:00 High-altitude Formation
Regensburg, Germany

Called for 0130 briefing to go after an assembly plant southeast of Regensburg, Germany. Flew #2 of the low squadron low group until we were forced to leave the formation. As we crossed the enemy coast I saw my first B-17 go down in a spin as the antiaircraft fire over Rotterdam opened up on a group to our left. This wasn't a pleasant sight and no parachutes came out. The formation scattered all over hell as the leader climbed at 155 and 160 mph, and we were only five ships of an element about 5 miles behind the wing formation.

Then, just as we pulled back in wing formation, we passed over our third flak area at Stuttgart where we got the hell shot out of our ship. Lt. McCann was hit, Sgt. Ackerman was grazed, #3 and #4 superchargers were shot out, our left aileron had a large hole in it, and our right wing gas tanks had nice holes through them.

B-17 taking flak while its bomb bay doors open and make ready to begin dropping its payload. Photo from the Harry Druccilli, Jr. Collection [Acc. # 2007.0485] / National Museum of the Mighty Eighth Air Force, Pooler, Georgia.

McCann had a hole through his right foot, which looked rather bad, but I had it bandaged up and applied with plenty of sulfur powder to prevent infection. After attempting a shot of morphine and finding the tube frozen, I warmed one up and gave him a shot in the calf of the leg. He rested OK until taken out of the ship.

I went back to the cockpit to find we had dumped our bomb load of 12x500 pound incendiary before reaching the IP* and were trailing the formation. We dropped to 16,000 ft. and flew below the formation to the enemy coast. Our gas was really low and when we landed we had about 20 gallons per tank. We had a P-47** escort from the enemy coast in, and they certainly looked good. The leader was flying "Boston Bull," UJ-M call letters.

*Initial Point is an identifiable landmark about 20 minutes from the target. From there on, the formation had to fly straight and level with the bomb bay doors open.

**P-47s were fighters.

(In an interview many years later, Daddy said that he had secretly purchased a thick, heavy piece of metal to place under the pillow of his seat in the plane. The piece of flak that knocked out the hydraulic system located under his seat would probably have hit him too if not for the piece of metal he was sitting on.)

Saturday, July 22nd
Practice Formation at 10,000

Was I ever pissed off when they called us at 0700 to fly practice formation. It's the lead crews that need practice, not us wingmen. We landed at 1300 and had to eat standing after they had taken all of the chairs and tables out of the mess hall. Caught a ride over to the 231 General Hospital and saw Lt. McCann. He received the package I sent him this morning; it was shaving stuff, stationery, etc. Didn't stay long because I made it back for the banquet held in the hangar to start our One-Hundred Missions party. General Doolittle and other big shots were there and a camp show was good entertainment. I had several drinks at the party later and got to feeling plenty good. But the way every officer propositioned all the women was downright disgusting. It looked like a bunch of dogs, like a bitch in heat, so I left.

Sunday, July 23rd

Slept until 1100 this morning and didn't feel the effects of last night at all. Spent a very quiet day around the barracks writing letters and batting the breeze. I hear some tall tales about last night's party and all these English women are still around, shacking up in the huts with different officers.

Monday, July 24th
Logged 6:30 Medium-altitude Formation
Saint-Lô, France

Called at 0630 for briefing on a mission to support the ground troops in their effort to capture Saint-Lô. We were to bomb in a rectangular area west of the town with the entire 8th Air Force taking part to allow the infantry to go through. We were loaded with 42x100 pound GPs and were to knock out the German main lines. Clouds prevented our bombing, so we brought the bombs home. Saw a B-17 in the wing ahead of us explode in mid-air. Flew #3 low squadron, high group. Coming home, the entire wing screwed up and our instrument descent was a hell of a mess. Picked up only one flak hole today. Ate a late lunch, so didn't get up for supper tonight and I am turning in early, for I fly again tomorrow. Don't have time to do anything these days but eat, sleep, and fly.

Tuesday, July 25th
Logged 6:35
Medium-altitude Formation
Saint-Lô, France

Called at 0300 for briefing on the same mission we took yesterday and were off at 0500. Today the ground troops are marking the front lines with red flares so we don't bomb them. Yesterday some dumb jerks dropped PFF and hit our troops. Saw several troop carriers in the channel and landing strips and hospitals in the beachhead. We were at 12,000 ft. and the battlefield was clearly visible. Our load of 42x100 GPs went off on time and our area was certainly well hit. The bombs exploding were a beautiful sight and I saw one hit a small farmhouse and it opened like a paper box. Never saw so many airplanes in such a small area, and with visibility cut to about 3 miles by a haze, it was a sweating job. It was a rough ride coming home, for after dropping to 3000 ft. the air was awfully rough and bumpy. After arriving back at the base I went to Norwich and had Stone take my measurements for a battle jacket, which is supposed to be ready on August 20th. Left a roll of film to be developed, which takes four weeks and costs a little better than a pound for the 36 prints.

Wednesday, July 26th

Was a groundhog all day and it felt awfully good to have a day's rest. All pilots were required to read a booklet on gas conservation in a B-17 and, from the way it read, the generals are getting tired of crews throwing guns, etc., overboard. It said that by ditching the ball turret and everything else possible for a return trip, your range is increased by only 23 miles. What a crock. If the weight doesn't mean so much, why is the 8th Air Force order out to not put any armor plating or bulletproof glass in our ships? Some of them should have to ride these missions. I shot skeet from a moving platform and only broke 2 out of 8. Went over to the 231st Station Hospital to see Lt. McCann tonight and he seemed fine. A Lieutenant just back from Normandy told me every house over there is an arsenal, for they are all made of rock.

Thursday, July 27th

Called at 0030 for a briefing on a trip to Stuttgart. But just as we started our engines the tower fired a red flare and the mission was scrubbed. Tried to cook up another mission, but at 0630 they called things off and I hit the sack to sleep until noon. Lt. Crawford kept me awake all night and it was a relief when this was scrubbed. Spent the afternoon lying around doing almost nothing as the rain fell for the first stormy day since I reached England. All our enlisted men received the Air Medal* at a ceremony tonight from our new group commander, Colonel Smith, who replaced Colonel Odom on July 25th. I should get mine sometime before long.

*Bomber crews in the ETO received the Air Medal after five combat missions.

6

FIGHTERS AT 12:00

Friday, July 28th
Logged 9:00 High-altitude Formation
Leipzig, Germany

Called at 0030 for briefing and we were after a synthetic oil plant at Merseburg, about five miles west of Leipzig. The mission over to the IP was perfect with no flak, for a change. At the IP, the lead and deputy lead attempted to change lead and when they messed up, I saw my first mid-air collision; both ships were PFF. One ship broke in half at the rear entrance. The other went into a flat spin and later broke up. After the tail broke off, the lead ship moved up, did a hammerhead stall, and went straight down.

Saw one man falling from this without a parachute and as the tail section went by, I saw the tail gunner come out and open his chute. Parts flew by us, for we were flying #2 low squadron lead group. But, luckily, everyone else escaped without damage. The crew was rather spirited and our bombardier, Lt. Oswald, salvoed* our bombs as he opened the bomb door. Our squadron stayed intact and took over the lead of the 45th wing coming home. In the mess-up, some failed to drop their bombs and were dumping them all the way back across Germany.

We lost our new CO, Colonel Smith, who was on his second mission since taking over last Monday. The two crews were well along with their missions and there was lots of rank aboard: a major, four captains, and the other officers were all first lieutenants. The pilots were Captains Bayer and Franklin. This was sure a mess. Over the target the flak was modest and inaccurate. Our load was 10x500 pound GPs. Engine #1 started throwing oil and after getting out over the North Sea, we feathered it and came in on three. Didn't pick up any flak damage. The food at dinner was piss poor and I don't mean maybe, though it may have been the way I felt. Slept all afternoon. Looks like I go out again tomorrow.

*A simultaneous or successive discharge of the bombs.

(As you can see, Daddy and most of the others in battle did not allow themselves to express their fear and grief. Saying the food wasn't good may have been because of the way he felt was about as much sadness as he admitted in all of this diary. In "The Mighty Eighth" by Gerald Astor, one of the airmen said that the crews generally stuck together and did not make friends outside of their crew. This was in self-defense. The fewer people you knew, the less grief you suffered when a plane was shot down.)

Saturday, July 29th

Woke up at 0900 to find my crew was out on a mission with a new pilot making his first trip as copilot on our crew. They went back to Merseburg and from what I heard they did a good job, so we won't have to go back. I am now back with the crew, for we all have 15 missions while Sgts. Ackerman and Perry have 16 each. It's rained here all day long and now it's getting rather cold. Finally found some heavy wrapping paper and got the package off to Mary Liz. Also mailed Elizabeth the package of coins. The boys burned out #3 today and the prop ran away, but only picked up a few holes, landed OK. I laid around the barracks and wrote letters. Bought three magazines today: Colliers June 17th, Liberty July 1st, and Cosmopolitan. These magazines are really up to date.

Sunday, July 30th

A stand-down* today, so I slept until almost noon and it certainly felt good to lay in the bed and hear the rain falling on this tin roof. Had an hour of ground school covering engine operation this afternoon. Saw a fair ballgame that didn't end until 2000 and it was still plenty light. Have another pass starting tomorrow; I tried to get off today but couldn't. Would have gone over to see Ben tonight, but tomorrow there is a maximum effort and we may have to fly.

*A temporary halt to offensive military action.

Monday, July 31st

Took off today on a forty-eight-hour pass to London, where I am despite all the buzz bombs. Coming in on the train I talked with a fairly good-looking English babe. She showed me around the town and after dinner we went to a show.

Later we were down on Piccadilly Circus, and there really was a circus. One blond wench came by telling another babe she couldn't get £5. This tickled me and we both laughed. The blond then turned to the babe I had and told her to mind her own business, and if she wanted to give her tail away for love, that was OK, but she was getting £5 for her ass. I took the gal to her home and didn't even kiss her goodnight, for I am afraid of these English babes.

Tuesday, August 1st

Spent the day walking around the town looking for some 8th AF gold-embroidered shoulder patches, but had no luck. Saw where a good many buzz bombs hit and the actual damage isn't so bad, but the concussions pushed in all the windows and doors for about four blocks around. Am staying in the Red Cross Reindeer Club*. Can't eat English food.

Went to the bar at the Regent Palace Hotel and had a few mixed drinks as I talked with all the Piccadilly Commandos** who knew nothing but £8. Several times I said, "Well, let's go up and go to bed." At this, their faces would always light up, for they thought they had a sucker. Then I would laugh and tell them no, and such long disappointed faces this would produce.

Ran into Lt. John Wincey from Live Oak, whom I met back in San Antonio almost two years ago. Went to the show and saw "Pin-Up Girl" for the second time and it was still fairly good. Also ran into Capt. McRae, who came over with the Walker Provisional Group. He finished his tour and is now flying a P51 weather ship over the target. This is the only one I know of still around and Lt. Bailey, a classmate of mine, is his copilot.

Ran into my enlisted men and found out our pass just started today, so I managed an extra day in town in taking off early. We watched the Piccadilly Commandos work for an hour and I must say these English have their nerve, for they walk right up to a man on the street and ask £8 to spend the night with him. Boy, what a price and what nerve.

*The Reindeer Club was the Officers hotel and club

run by the American Red Cross.

**Prostitutes.

Wednesday, August 2nd

Took a tour around London to see all the points of interest. The guide showed us St. Paul's Cathedral, Westminster Abby, the Houses of Parliament, Buckingham Palace, #10 Downing Street, London Tower, etc. The guide was a French refugee and she knew all the dirty parts of history and really gave an interesting tour. We rode in one of these funny-looking taxi cabs with the top down. Bought a guide to England and a map to record my missions on. Tired, so I am turning in early tonight.

Thursday, August 3rd

Slept rather late and would have slept more, but the sirens kept blowing air raid alerts. Walked around looking the town over and hunting a cable release for my camera. Bought Mama some flower seeds from a seed store. The buzz bombs came over all night and morning, with lots of them right overhead, and quite a few explosions were nearby. Caught the train out at 1400, getting back to base at 1900. Had lots of mail and a package from Mary Liz, so we all ate pecans, cookies, and candy — they are really good. Lt. Crawford and the others moved out today and Lts. Martin and Seline have gone to PFF, so Haines and I now have the place to ourselves.

Friday, August 4th
Logged 6:35 High-altitude Formation (24,000)
Bremen, Germany

Called for briefing at 0415 and found we were going after an oil works in Bremen and, after a one-hour delay to allow a heavy fog to lift, we took off on our journey. We flew lead ship low squadron of low group, but this was an easy place to fly. Though we were over enemy territory only twenty-seven minutes, there was plenty of hell. We picked up quite a few holes and I am surprised we didn't get more, for this flak was the most significant burst I have yet seen. Saw three B-17s go down in flames, but in each case at least five parachutes came out of the ship.

Lt. Awswell, our bombardier, really pissed me off when he dumped our bombs about two minutes before reaching the target and then started yelling over the interphone to drop out of formation and not go over the flak area, but we went right on through and our lead ship was hit bad and his radio operator was killed. We lost one ship today over the target, but I didn't see him go down. The Jerries threw up three bursts of flak, which burned with a light red smoke; these were formerly used to signal German planes that our formations were out of flak range and they could come on in and look things over. Our squadron was well hit and many of the boys came home with feathered props and flat tires. Red flares were numerous at landing time. No enemy fighters showed up.

Tonight I found out the officers of our crew are on mail censoring detail. After reading through a mess of letters I stopped reading, just opened them and signed them as OK if they were written in English. The mess hall wouldn't feed us after the mission until supper, which made it twelve hours between meals.

Saturday, August 5th
Logged 7:30
High-altitude Formation (24,000)
Magdeburg

Called at 0430 for briefing and went to Magdeburg after a big iron manufacturing industry, which puts out Mach IV tanks and antiaircraft guns. Saw lots of dogfights off our right wing on the trip in. The Germans have some type of smoke bomb under their wing, which is set off as our fighters fire at them. When the smoke starts pouring out, they go into a tight spiral and down. But this doesn't fool our fighters, for they follow them down. I saw one Jerry pull out down around two or three thousand and two

P51s were right there and shot him down.

Aftermath of a bombing mission. Photo from the Irwin Math Collection [Acc. # 2008.0151] / National Museum of the Mighty Eighth Air Force, Pooler, Georgia.

Saw our first jet-propelled (rocket) fighter today at a distance; though they didn't attack, we could see they were faster than hell. As we watched and broke off in groups for the turn at the IP, a formation of 12 to 15 Me109s pulled up from somewhere at 10:00, ready to make a pass. I couldn't make out what they were, but told the boys to keep an eye on them. They pulled on ahead, made a turn, and came in for a head-on attack and as they turned, I called, "Fighters at 12:00," and our bombardier and top turret got in several bursts at the one nearest us. The few seconds of this pass were the longest of my life as the noses of all these fighters spouted flames as their guns blazed at us. All the Forts came through OK and our escort gave the Germans a good chase.

Before this the formation was loose and very few gunners were in their turrets. Afterward the formation was tight and all turrets were manned. The flak was heavy but not accurate and our group came through OK. As we turned off the bomb run I saw three ships of another group go down; two of these exploded as the flak hit them and went down in small pieces. We flew lead ship of the high squadron low group with a bomb load of 10x500 pound GPs. Had my second package from my lovely wife with some eats and other things. Had a critique on the last seven missions tonight and found out I missed two short missions while on pass.

Sunday, August 6th

Didn't fly, but the boys went to a target down near Berlin. The crew shot skeet and I finally had a good day getting 20 of the 25 birds. Collected my July pay, £30 ($121.00), and bought a $50.00 war bond to help the 8th Air Force Victory Squadron drive. Our squadron CO, Lt. Col. Lyles, called a meeting of all 731st officers and gave us a little talk. Went over to see Lt. McCann for a couple of hours. Had a package from Sam Jr.* and the gum and razor blades will come in plenty handy.

*Daddy's brother.

Monday, August 7th
Alt. Formation (20,000)
Reims, France

Called at 0500 and briefed to take out a double-tracked railroad bridge about four miles west of Reims. Had a load of 6x1000 pound GPs, but after making one run over the target under a 10/10 overcast* we returned to base with our full bomb load, for the PFF couldn't pick up anything. We hit flak going in and coming out. Some were really close and one went off about ten feet above our right wing. This really jolted us but we only picked up three holes, none really bad. Had two packages from my wife tonight and the eats and things were delicious. Finally received my Air Medal tonight after finishing eighteen missions. Heard my recommendation went in today for a promotion, which should come in before too long now.

*Total overcast.

Thursday, August 8th

My crew flew a milk run, taking off about 1000 and going in support of the British ground forces; they were back shortly. Some new pilot flew in my place, so I am now one mission behind the rest of my crew. They saw only a few bursts of flak. Went to Norwich to see about my battle jacket but Stone was on his two-week summer vacation, so I didn't stay long. Spent an hour looking through the Norwich museum. Rather pissed off at missing this mission. The war news looks good and I still think we will leave England before September 15th, and the war will end about then.

Wednesday, August 9th
Logged 5:30 Alt. Formation (21,000)
Nuremberg, Germany

Called at 0400 and briefed to go to Nuremberg after a diesel engine plant with 10x500 pound incendiary clusters and were flying lead ship low squad of low group. Due to a frontal system, we descended from 23,000 to 18,000, and some dumb navigator took us over Liege, Belgium. When the flak opened up, our group lead made a 180 degree turn around the flak hub and they really gave us the works. We lost two ships out of our formation, one in occupied territory, the other made a successful ditching. We had our hydraulic system shot up and it went out on landing, so we rolled off the end of the runway, made a slow ground loop and stopped with no further damage. The hydraulic system was shot out right under my seat. A piece of flak missed me by only inches, the ship had twelve flak holes.

Thursday, August 10th

Slept until 0900, went to two classes in ground school and ate dinner. Caught a ride down to Bury St. Edmunds to see Frank Farnbach* then found out the 92nd group wasn't near there. Came back by 3rd Division Headquarters, which is in some Lord Ivey's party house southwest of Thetford. What any man could want with a house like this is beyond me, for it looked like a big hotel. His real home was down the road about half a mile. This estate covers about seven or eight square miles of land where this fellow has his private hunting club and the king comes up there for hunting each season. This is a beautiful estate and he even has his own private church and cemetery. Out at the 96th group, I saw what was left of a B-17 after it attempted moving a tree. What a mess that was.

*Frank Farnbach was also from McIntosh.

Friday, August 11th

Missed an easy mission, for they went after a marshaling yard* in southern France and found only four guns on the target. Found Lt. Alex Bebenin's (our old bombardier) new address and am trying to get his baggage off to him. Jack Bateman** was up to see me this afternoon and after shooting the bull about home and other things for two hours, he took off in his P47 giving the field a good buzz job before heading for home. Turned in a copy of my active duty orders today to go with a recommendation for my promotion to First Lieutenant. That extra dough will certainly come in handy.

*Where trains are assembled.

**Jack Bateman was also from McIntosh.

Saturday, August 12th

Called at 0500 for a practice mission and everyone was bitching. I didn't have to fly, for the operations officer took my place. They were practicing a new five-group wing formation. Packed Bebenin's stuff and have it ready to go, but I see signs of another mission to Russia as they bring in fighter belly tanks. Went over to see Lt. McCann and enjoyed talking with the men evacuated from Normandy. Going to turn in early tonight.

Sunday, August 13th
Logged 6:30 Medium-altitude (18,000)
Nantes, France
Load 42x100 pound G.P.

Called at 0600 and was rushed, so we didn't get breakfast as we went to support the Canadians in Normandy. Saw very little flak and several rockets as we bombed in three and four ship elements for the first time in the 8th Air Force. Our formation came home alone, for we were the lead ship and couldn't find our group leader. Picked up five small flak holes. Went over to the 388th bomb group near Thetford to see Lt. Col. Ben McLaughlin and Lt. Jack Bateman, and with lots of other Florida boys, we had a big time closing the party by singing the University of Florida song. Saw a lot of my old friends from training days and heard of the ones who were lost. Ben sent me home in a car, for I will probably fly tomorrow.

Monday, August 14th

Surprised when I woke up at 1000 and found I had been left off the mission, and this time it was southern Germany. Relieved Lt. Haines as duty pilot* at lunch. Went to Norwich to see Stone, and the material for my battle jacket just arrived. He will fit it on Friday. Bought an 8th Air Force gold-beaded patch to send to my wife. Stopped by to see Lt. McCann on my way back. The harvest here is now in full swing and many of the grain fields have already been cut. Ate some Vienna sausage, tomato juice, and cookies for supper. The crew that our group lost going to Russia on June 21st came back today, as seven of them evaded the Germans after bailing out near Warsaw and getting back through the Russian lines.

*The designated officer who handles the non-flying duties such as scheduling, assigning duties, handling issues, etc.

Tuesday, August 15th

Slept through breakfast. Went down to the photography shop and picked up some pictures of our crew and formation. Took a typhus booster shot and my arm was plenty sore for several hours. Got two weeks' rations. Our crew is alerted to make another shuttle run to Russia, which was scheduled for tomorrow, but has been postponed. Looks like we may help the infantry again tomorrow. The Allied landing in Southern France and all other war news tonight looks good for an early victory in the ETO. I say before September 15th.

Wednesday, August 16th

Gas mask lecture and gas chamber drill this morning and, as usual, it was a mess and a crock. Found out my mask was damaged by the insect repellant spilled in it on the Russian trip, so I drew another one. Picked up a few formation pictures from the photography shop. Collected and took Lt. McCann his July pay. Saw quite a few pheasants in the grain fields, which have been cut. Rode over to the hospital with Capt. Wright, our flight surgeon. Shot the breeze around the club for a couple of hours after supper, came home, wrote a few letters and went to bed.

Thursday, August 17th

Messed around all morning doing odd things and picked up another light flying jacket. Went to Norwich only to find all the stores closed for a half-day holiday, which they take every Thursday. Stopped by to see Lt. McCann and took a picture of him inside the hospital ward. Haven't heard any more about my promotion, but I suppose it will come through before too long now. We have another stand-down tomorrow. And I hear we aren't flying any more operational missions before going to Russia. Hope we hurry up and get moving again, for I want to finish up this mess. Two P-38s really gave us a good buzz job and acrobatic show just before dark tonight.

Friday, August 18th

Got up for breakfast for a change today and messed around the base doing almost nothing until noon. Went to Norwich and had my battle jacket fitted; it will be ready Saturday week. Found a pair of embroidered wings, so I will send all this stuff to Mary Liz tomorrow. Got my first roll of pictures back today and they are fair. Took four weeks to get them back. Looks like the Russian deal has fallen through, so we may not go back. Today's war news is excellent as we move on in France, but these damn English aren't doing a damn thing down there. They should have their asses kicked for taking so much credit. We should be moving before too long now. My promotion is dated August 17th if it ever comes through.

Saturday, August 19th

Sgt. Lundstrum and I shot skeet and I managed to get 18 of the 25 clay birds. On this new range they really come out fast. Played volleyball for about two hours, and tomorrow I expect to be too sore to move. Stand down until 0830 tomorrow. This group will be moving from the looks of things, for every department is collecting shipping boxes. The war news looks good and our job in the ETO is about finished.

Sunday, August 20th

Sacked until noon and attended a two-hour class on automatic pilot after a lunch of hot dogs and potatoes. Cloudy all day with a drizzle, which was just enough to get good and damp. Went to the party at the club tonight, and after several drinks and a couple of dances, I came home, for most of the boys were making regular asses of themselves. There were three or four fair-looking girls, and everyone was after them. Saw one young kid leave with some old English hag about 35 and ugly as hell. Looks like these boys could be a little more refined with their doings.

Monday, August 21st

Sacked until noon and after eating a good steak dinner (small pieces of steak), I went over to the hospital to see Lt. McCann. Rained all day today and everything is damp and sticky and plenty cold. The war news tonight is really good and I may not be here too much longer. Though our Russian deal seems to be off, something is brewing and I feel this outfit will move sometime soon. Every outfit is busy getting packing boxes ready to move on short notice when the order comes through. Heard the bomb dump is about empty, and the other day I saw them hauling some bombs away from this base. Looks like I will see some more of the world in the near future.

Tuesday, August 22nd

Shot skeet this morning in a high wind with fast traps and picked up 18 out of the 25 shots. Had a class in ground school for two hours and a meeting at the club for all officers. We spent $451.00 paying for a jeep wrecked here last February 20th by two nurses after a big party; that was before my crew even arrived here. Last night we had a red alert but I just turned over and went right back to sleep. It rained all night and today there has been a damp mist falling; everything is grounded. Got a piece of Plexiglas to make a letter opener and give me something to do in my spare time. Received a carton of gum and a sponge to saddle soap my shoes from Sam, Jr. today.

Wednesday, August 23rd

Called for a mission at 0900, but before getting to the briefing room the mission was scrubbed. Had two lectures, one medical job and another by First Officer Stein. He is one of our boys who was shot up and forced to bail out over Normandy; he told us his experiences. Shot skeet and only broke 16 of the 25 shots. Bought my week's rations. Lt. E.G. Oswell, our bombardier who was so flak-happy, left today for the States after they let him off with 30 missions. He sure was a happy fellow. Sent Mary Liz some papers and two Yank magazines today. News tonight sounds good, for Paris was liberated today, which was the biggest news of the war in France. Every news program told almost nothing but this story.

Thursday, August 24th
Logged 1:15
Practice Wing Assembly Route

Flew copilot for a 729 Squadron crew taking lead ship training and had to fly the assembly group; this pilot sure couldn't hold a constant airspeed. Rained all afternoon, so I stayed in and wrote a couple of letters and shined some shoes to pass the time. We were alerted for the shuttle run to Russia but it was scrubbed and now we are flying a regular mission tomorrow. The war news looks better every day and now I may not get my thirty missions for the DFC* before this mess ends in the ETO. Really, I don't care, for I would love to start home tomorrow.

*Distinguished Flying Cross — by the time he received his, the required number of missions was 35.

Friday, August 25th
Logged 9:15
(24,000 Formation)
Stetten, Germany

Called at 0430 and briefed to go after an oil refinery in Stetten, Germany, which we hit hard. This was certainly a long haul and I am plenty tired now. No fighters, flak at target was heavy barrage with some tracking. We took an 88 shell through the left wing tip, which didn't explode, lucky for us, but left a rather large hole. Had several small pieces through the ship and one took the #4 engine out; we feathered, couldn't stay in formation so unfeathered, and ran her rough all the way in. Found out on the ground that a broken tie rod was causing all the trouble. Our group was well shot up and there were quite a few wounded men today. Two ships came in with hydraulic lines shot out, one landed OK, but the other wrecked, so salvage will take care of the rest. Saw rockets launched at two planes, but they were inaccurate. Major Glen Miller and his orchestra are playing over at the 388th Bomb Group. But I am too tired to go, even for this. Returning across Germany I saw several good fires where other wings ahead of ours had hit oil targets. Upon returning to the base, this traffic pattern is a regular rat race, worse than back at Basic. We flew #2 lead squadron, low group, and our bomb load was 10x 500 pound GP. Due to weather, we assembled at 500 ft. and went out over the North Sea before starting our climb.

Saturday, August 26th

The group pulled another milk run to Brest. And here we sat on the ground; they didn't see fighters or flak. Cut ground school and took life easy this morning. Went to Norwich and picked up my battle jacket. It's really nice but not worth the £13 that Stone soaked me for ($52.00). Found a golden wedding card and mailed it to Grandmammy and Grandpappy. Hope it reaches them by September 4th. Ate supper at the Red Cross and returned to base rather early. Took my rationed shower and am now nice and clean and ready for a good night's sleep.

Sunday, August 27th

Wrote letters all morning, ate an early chow, and caught a ride to Cambridge. Secured a room at the Red Cross and tried to call Lt. Kirby. Ran into Lt. Bradner of the old Walker Provisional Group, who has put in his missions and now holds a ground job at the 351st Bomber Group. He was in charge of a convoy taking girls out, so I went along with them to see Kirby and we two spent the evening shooting the bull. Caught the same bus back and was about raped before getting off. These English girls are certainly hard up for men, but I am not interested. From this load, two asked me to go home with them. Made the Red Cross shortly after midnight.

Monday, August 28th

After catching the bus and several rides out of Cambridge, I finally found Podington and the 92nd group, and then it was no trouble finding Corporal Frank Farnbach. Have never seen anyone the Army did more good for physically. He really looks healthy now. Enjoyed the four hours I spent talking with him. Caught the bus back to Cambridge and a ride on to base, getting here about 0100. Travel by train or bus stops at dark and ride catching isn't so good then either. Saw the Eighth Air Force cemetery outside Cambridge today, for it's here that everyone killed in England is buried.

Tuesday, August 29th

Had a nice package from my wife today and the cookies she sent are just what the doctor ordered. Went down to Ipswich to see Sgt. Harry Terrell but upon calling the base, I found he was MIA*. Didn't have enough time to get out there, so I am going back before long and find out when and how he went down. Caught a ride back to Norwich, ate supper at the Red Cross, took in a stage show and caught the bus home. Saw a lot of England on this 48-hour pass, had fun, and didn't spend much money.

*Harry Terrell was a friend of Daddy's from Ocala. He was shot down over Belgium and walked out of the country with help from the underground.

7

NOW A FIRST LIEUTENANT

Wednesday, August 30th

My promotion orders came through today and they are dated August 24th, so after 13 months as a shavetail*, I am now a First Lieutenant. Went into Norwich to get some bars but when the Limeys asked 9 shillings per pair, I decided I wasn't a sucker. Bought a couple pounds of apples and this fresh fruit certainly tastes good. Priced a peach, and they are 4 shillings each.

Walked into the club tonight and ran right into Emmett Clonts from back in our Stetson** days. He is now a major with our group just coming over from the States. Small world, isn't it? The group went to Bremen, Germany today, bombed PFF, and had light flak with no fighters. Major Gill is now our squadron CO, replacing Lt. Col. Lyles.

*Shavetail is a newly commissioned officer or one with little experience.

**Daddy graduated from Stetson University in DeLand, Florida.

Thursday, August 31st
Logged 3:00 Local Flying

Slept through breakfast as some of the boys went out on a practice mission this morning. Someone buzzed our ship and when the two met in mid-air, the horizontal stabilizer was damaged, so the ship is out until they get that fixed. Had ground school scheduled this afternoon when called out to test hop a ship. Were supposed to take 909 to altitude, but these engines didn't have enough time on them, so we just slow timed* it around the local area. All the enlisted men tried their hand at the pilot's controls. I made my first takeoff and landing in over six months and my third since finishing cadets. Rumors are making the rounds that we may go back to the States for B-29s**, which would be a good lick.

*To slow time an engine is to allow a new or rebuilt engine a "break-in" period.

**The Boeing B-29 Superfortress was flown primarily

by the United States during World War II and the Korean War.

Friday, September 1st
Logged 8:15 Alt. Formation (20,000)
Mainz, Germany

Called at 0245 and briefed to go after a truck factory in Mainz, southwest of Frankfurt, Germany. We flew #2 lead squadron, low group, and carried 4x1000 pound GP and 4x500 incendiaries. Flew through an overcast for almost an hour; we had a recall when only about 20 minutes from IP. Coming back, we flew down the Seine River at 10,000 feet, could see all the blown-out bridges and the allied pontoon bridges replacing them. Saw Paris and the famous tower and big auto racetrack. Saw lots of battle damage around towns and airfields. In the English Channel, I saw the entire Allied fleet; at least that's the way it looked. Brought our bombs back. Rode a bike over to see Lt. McCann at the hospital tonight. Think we will get credit for this mission, making mine 22 with 13 more to go.

Saturday, September 2nd

Slept through breakfast and just stayed in the sack until noon, for it was raining all morning. Went into Norwich to get a pair of pants I had left with Stone to have altered. Also picked up some wings for our gunners. The weather was so nasty that I came on home early. I cut ground school this afternoon. Hank Haines told me tonight he will be made a command pilot and I am supposed to be checked out as first pilot and take the crew. The carton of gum from Wilshire* finally came through today and now I have almost three cartons of gum.

*Wilshire was Daddy's older brother.

Sunday, September 3rd

The boys made another milk run to Brest, France, and, as usual, here I sit on the ground. The lead ship screwed up and went over Guernsey Isle and antiaircraft shot it down. We lost Lt. Col. Lyles and Major Clonts in this ship. This was Emmett Clonts' first mission and as near as I can learn, they probably got out OK*. I was supposed to be checked out as first pilot today, but they called off the flight. Took in the party in big shape tonight and after getting rather high, I had a good time dancing with these English girls. Everyone was there and we really had a good time until everyone got a little too much to drink and then the fights started. I just stood back and watched them tear into each other.

*Emmett Clonts was the friend from Stetson that Daddy ran into on August 30th. He was their flight surgeon and on his first mission. He died that day.

Monday, September 4th

Haines is now squadron grounds school officer and I am almost sure of being checked out, but the weather today had all ships grounded. Played poker as it rained all afternoon and won $10. It's still falling on this tin roof and I am certainly going to use it for some good sleeping.

Tuesday, September 5th

Went down and shot skeet this morning with the crew and only broke 16 of the 25, for the high gusty wind had the birds doing funny tricks. Was supposed to be checked out today but all ships were grounded, as some change was being made on the life rafts. The group was on a mission down near Frankfurt somewhere.

Made silver bars for my leather jacket by gluing silver from chaff over the old gold ones. Tonight I bought a good bicycle from a lieutenant who came over with us from Sioux City and is going home on 28 missions. It's really a honey and cost me £8; I feel sure I can sell it for almost that much when I leave.

Wednesday, September 6th
Logged 3:00 Local Flying at 8,000 ft.

Called for briefing at 0230 and were scheduled to bomb Berlin. Had a time delay and another briefing, then this mission was scrubbed. At 1030 we had another briefing and this time it was Bremen, but shortly after takeoff this was scrubbed. We flew three hours to burn gas and lighten the load for landing. Were supposed to fly formation but everyone started flubbing the dub* and shooting off over the radio like a bunch of dumb jerks and no one in our group ever got in formation.

*The dictionary says "flubbing the dub" means to loaf or evade duty or bungle. It also says this term was widely used during WWII.

Thursday, September 7th

Rained all day and I certainly enjoyed staying in bed until about noon. Went to the photography shop and picked up four good combat photos. Went over to see about getting checked out but all ships are grounded and we couldn't shoot practice landings. Caught the ambulance and rode over to the hospital to see Lt. McCann, who will be out of the hospital before too long now. Stole some coal* and now we have a nice fire going in our stove, but it's almost too small to heat the tin shack. To hold the heat in, we put paper over the ventilation on each end of the barracks.

*They were supplied with wood to burn, but it didn't heat well. The coal kept them warmer. There was a coal pile near the HQ building.

Friday, September 8th
Logged 8:00 Alt. Formation 28,000
Mainz, Germany

Called at 1230 and briefed to go after a tank factory but due to PFF bombing, we were forced to bomb a supply depot across the river from Mainz. Flew copilot of Lt. Gardner's crew on their first mission. We flew #4 low squadron high group with 10x500 pound GP. There was a ground fog, so we had an instrument takeoff and, due to the overcast, we didn't break through until we reached 14,000 ft. Coming up, we picked up lots of ice on the wings. I flew quite a bit of the formation and just before the IP, when I attempted to take over #6 of the lead, I had the scare of my tour.

The low leader and #5 lead pulled a squeeze play; I cut the throttle and managed to pull out and get back into formation for bombs away after falling through prop wash. We missed a mid-air collision by only a few feet. Flak was light and there were no enemy fighters. At one time the flak was tracking rather light, so we pulled out in evasive action and four bursts went off right close to where we would have been. The entire trip was above an overcast. The temperature was -44 degrees centigrade and one of our boys had an orange freeze and burst in his pocket. The cockpit's overhead windows froze over and frosted up, so we couldn't see out at all. Slept about two hours before supper tonight, at which time we had some fair steak. Stole some more coal and am now sitting around a good hot fire just like back home; everyone here is currently shooting the bull and having a good time. Ate oranges and cookies and am now full and ready for bed. Tonight we had a squadron officers meeting and Major Gill, our CO, gave us a little pep talk.

Saturday, September 9th
Logged 1:00
Practice Takeoffs and Landings

Today the group pulled a short-haul to the western edge of the Rhine Valley and though the group lost no one, there were nine aborts and after going over the target, the others brought the eggs* home. Was out shooting some landings with Haines this morning getting checked out when on the third landing, which was the best of the three, the right wing dropped and as the ship slowed down, it finally gave, and we ground looped off to the right, dragging the wingtip and #3 and #4 engines. When I looked, I found the drag strut had broken, allowing the landing gear to collapse. Went to Operations and made out an accident report. Spent all afternoon in the sack and tonight we have been shooting the bull.

*Bombs.

Sunday, September 10th
Logged 8:30 Alt. Formation 25,000 ft
Nuremberg, Germany

Called for an 0330 briefing and sent after a tank factory southeast of Nuremberg with maximum gas load and 5x1000 pound GP. This was my first mission as pilot and I had our regular crew with Lt. Young, some smart kid, for a copilot. He pissed me off trying to fly so closely, so I changed to the right seat and flew most of the mission. The cabin heat went out before leaving England and in the -43 degree centigrade temperature, my hands felt frozen. Next time I will know to wear my heated clothing regardless. There were no fighters and the flak over the target was light and inaccurate. In several other places they fired flak at us, which also was never even close. Bombed PFF and hit the target. The takeoff today was piss poor as I wandered back and forth across the runway. For a while, we had a gas leak out of #1, but this finally stopped, then the carburetor air temperature gauge on #1 went out and lastly, the left-hand oxygen system went out, but we made it back OK. The landing was good but stopped at the end of the runway with a sudden jerk as I applied the brakes rather hard. Really pooped, so I went to bed and slept through supper. When I woke up, I had cookies, sausage, and juice my wife sent me from home. Sat around the stove and flew this mission all over again.

Monday, September 11th

Called to make a flight to Ireland, but as things worked out, I didn't get to go, for they already had both pilots. Went down and shot skeet, breaking only 14 of the 25 for a poor day and lost 5 shillings in a pot taken by the high man. Picked up another picture from the photography shop. Bought my rations. Caught the ambulance and took Lt. McCann some clothes and stuff he wanted. He should be out of the hospital before too long now. The EM* of the squadron had a party tonight and I have never seen so many drunks in one place in my life. Some joker stepped out and emptied his 45 into the air and there were fistfights galore. When these boys have a few drinks, they certainly go hog wild.

*EM means enlisted men, not the officers.

Tuesday, September 12th

Reported to Operations this morning and signed an accident report on the ground loop we did the other morning. Returned to meet the board this afternoon, which was short and sweet. Think I will hear little more about this matter. Went by the photography shop and picked up another picture. Those tight bastards won't give you more than one shot at a time. With nothing else to do this morning, I found a good broom and really gave our hut a sweeping, and when several of the boys returned, we decided to mop the place down. Played volleyball for a couple of hours tonight after supper.

Wednesday, September 13th
Logged 7:30 Formation at 26,000 Ft.
Bielefeld, Germany

Called at 0330 and briefed to go to Mainz, Germany after a sub depot. In our first pass the lead bombardier didn't pick up the target and a second run wasn't authorized. So instead we hit a marshaling yard at Bielefeld, Germany. Flew #4 of lead squadron, low group, and we unfolded* on the bomb run for the first time. My element slid out and the low group moved under for the bombing run. Our load was 6x1000 pound GP. Flak was moderate but accurate and we lost our #4 engine when flak cut our gas line on the tank and drained it. I sweated the entire time this 300 gallons of gas was running out the trailing edge of the wing. This happened just after bombs-away and we fell out of formation, coming home alone. Lt. Johns, a new pilot, rode with us as copilot on his first mission. When the flak bloomed up off our right wing, he moved over my way, but he moved back and took the middle area when he looked out of the left wing. One big burst out front made him duck when we went through the black cloud left by the explosion. Landed ahead of the formation, making my first three-engine** landing. After interrogation, I went right to bed, sleeping until a late supper and then a shower made me feel good.

*Moved out of the pre-planned formation.

**B-17s had four engines.

Thursday, September 14th

Called at 0300 and briefed to go to Berlin with a bomb load of 38x100 pound incendiaries and maximum gas load. First Officer Stock was supposed to fly as copilot and, in his excitement, he forgot his Mae West and had to return and get it. There was a heavy ground fog and, after delaying takeoff several times, they finally scrubbed the mission. We were really going to see what big B looked like today, for our target was almost in the center of town. Slept all afternoon, which helped matters. Am duty pilot now and down at Base Operations where I am on duty from 1800 hours today until 0700 tomorrow. Read my accident report tonight and it was written up as 100 percent material failure, which keeps my record clear. To do this job of duty pilot, all you really have to do is sit here.

Friday, September 15th

Only slept a couple of hours, sitting up in a chair, but was relieved at 0700 and went to breakfast, after which I went to bed and slept until almost noon. Shaved, took a shower, dressed in clean A's for dinner, and caught the bus to Norwich. Looked for some silver bars and tartan fabric but had no luck. Caught a ride to Ipswich, getting here rather late and took a room at the Red Cross Club. Tried to call Jack Bateman but he was up flying. Took in a stage show at the Hippodrome Theatre and it was fair. Went to the Red Cross dance and had a good time dancing to an American GI band's music. Ate a late snack and went to bed about midnight.

Saturday, September 16th

Got up rather early, went down to see the "Old House of Ipswich," which was built in 1576 or thereabout; all this was very interesting. Went to catch the bus to Debesh and found it had been discontinued, so I called and found out Harry Terrell was back in London. Ate lunch and caught the train to town, arriving too late to look up Harry tonight. Found a room at the Reindeer Club. Went to the Regent Palace Hotel bar, where I had a few drinks and met some of my crew's boys. Ate supper in a Spanish restaurant. Went pubbing until they closed. Stood around Piccadilly Circus until midnight watching the show and was propositioned thirty-nine times by women walking the streets asking from £5 up. Never saw anything like this before in my life.

Sunday, September 17th

Heard several rockets go off early this morning and one shortly after noon, which shook the ground where I was standing. Tried to find Harry Terrell through several army offices but couldn't, so I checked the Columbia Club* and found him registered there on a wild chance. But after looking all day, I had to catch the late train home. While waiting, I really enjoyed talking to several enlisted men who had just returned from France after bailing out over enemy territory. Had two letters from my wife and several snapshots, which I really enjoyed. The paratroopers landed all over Holland today in the most significant jump ever yet made and there is talk of this being another invasion.

*Columbia Club was a Red Cross club for the enlisted men.

Monday, September 18th

There was a stand-down today and I made full use of a good, late sleep. Have orders sending my entire crew to the flak farm* for a week and am now at Eynsham Hall, a big estate ten miles southeast of Oxford, having flown down. This is a beautiful three-story home on a 2200 acre estate of Lord and Lady Mayson. Here we eat five times a day: breakfast, dinner, tea, supper, and a midnight snack. Drew civilian clothes to relax and take life easy around this place. Went boating on the lake and took a short ride on a bike before dark. Played several different types of card games tonight. We are sleeping six men in a room in single beds in these huge bedrooms. The clean sheets and soft feather pillows certainly feel good. This is almost like home and certainly makes me dream of home.

*A flak farm or flak house was a place for R&R set up by the American Red Cross for combat weary airmen.

Tuesday, September 19th

Had fruit juice served in bed, then got up, ate breakfast, and caught the early train into London. After looking for Harry Terrell for about four hours, I finally found him at the Columbia Club. He looked good and told me all about the wonderful time he had during the thirty days he was down behind the German lines in Belgium. It was good talking to him, but messed around and missed my train.

Caught a later train to Oxford, where I stayed for three hours, then I called Eynsham Hall and had them send transportation for me. Had a big time looking the town over, for most of the street lights were on, and it certainly looked good to see a town lit up again.

Wednesday, September 20th

Slept late and that good sack really felt good just to lay there and take life easy. Rode a bicycle into Whitney, which is a nice, clean little town about five miles from here. It certainly felt good to wear these civilian clothes, and the bicycle ride wasn't bad. The country down here is beautiful with its rolling hills and is much nicer than North Anglia. Rode over to see some Roman ruins this afternoon, which were built about 1,200 years ago. Tonight we played cards and worked on a couple of jigsaw puzzles. Then we got together to pop some popcorn and shoot the bull, and the fighter boys told us some of the stories about their strafing* jobs.

*Strafing is attacking a target with a machine gun from a low-flying aircraft.

Thursday, September 21st

Cycled over to the 317th Street Hospital this morning but couldn't get any clothing coupons; did manage to get a can of orange juice. Ran into a maternity home and I have never seen so many pregnant women. After noon we cycled over to Blenheim Palace and looked over the place but couldn't talk the guard into letting us inside. This is a huge old messy-looking place built in 1705. Took quite a few pictures today. Went canoeing for a short time this afternoon. Had steak and ice cream for supper. Saw peaches and grapes growing in the greenhouse today, and it's no wonder things are so expensive. Saw pears growing on trees trained to grow on fences and apples on bushes growing on brick walls.

Friday, September 22nd

Slept rather late and then wrote my wife a letter. Took a few pictures of the house. Rode the bicycle into town and, while in Whitney, I had a shampoo; the hairdresser put me under an electric dryer and almost cooked my brains. Just beat the rain home, and it poured. Helped work a jigsaw puzzle and played cards until time to dress for dinner. Saw a movie but am now in bed as the second show gets underway. In one of the baths here, we have an all-wood bathtub, which is really a sport.

Saturday, September 23rd

Played three sets of tennis and won all of them. Don't know how, as I haven't played in over four years. After lunch, we rode bicycles down to the River Thames where we crossed a toll bridge and rode down on a footpath by the river for four miles. This was beautiful. The toll bridge was a scream, for there was a set charge on everything: horses, cows, oxen, sheep, and even chickens. Stopped along the river and gathered a few hazelnuts.

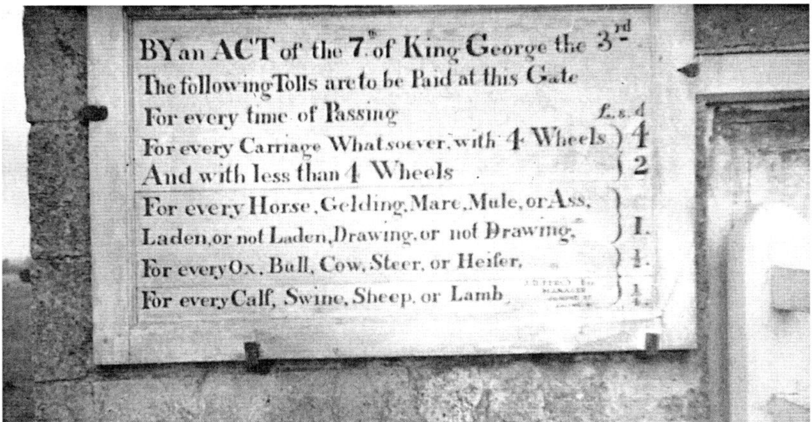

Daddy got a kick out of this sign he saw on a bridge over the River Thames.

Big dance tonight and for women we had a whole mess of Army nurses but not a good-looking one in the bunch. I got to feeling good after a few shots; some of the fellows went on real benders. Drinks were eighty cents for a single shot of scotch and Coke, so I didn't do too much drinking.

Sunday, September 24th

Today the weather has been cold and stormy with rain, so I didn't leave the house. Played cards and worked on a jigsaw puzzle besides reading to pass the time. Had several really good bull sessions. Packed my things to go back to the 452nd tomorrow morning.

Monday, September 25th

Took a good hot bath in one of the large tubs, ate breakfast, paid £3.10 for my week's board, and prepared to leave the flak house. Arrived in London and learned Harry Terrell had checked out and evidently started for home. Saw where Bill Bassett from Clearwater, Florida, had signed The Red Cross register and I got his APO number*. Saw two shows, "Double Indemnity" and "A Wing and A Prayer." Both were good but they each cost $1.00 in U.S. money. Had a couple of drinks and while sitting at the bar, Lt. Mitt Hauser came in, just returning from the States to start his second tour. Over a couple more drinks, we talked about the old Walker Group and how many of the boys are down. He saw Lt. Hutton in the States and found he had finished his tour in Italy.

*APO is an Army Post Office.

8

A REALLY GOOD BALL GAME

Tuesday, September 26th

Off to a bad start when a cabbie overcharged me for fare from the Reindeer Club to Liverpool Street Station. Train crowded with civilians going to races at Newmarket. Arrived at base in time for dinner and had lots of mail and packages. Lt. McCann is back from the hospital and gets around well on his bad foot. Lt. Ruckemena went down on yesterday's raid, and Lt. Col. Sherbourn, our new CO, was critically wounded but is expected to pull through. Corky, a roommate of mine, was hit in the leg and is now in the hospital. My enlisted men called today saying they will be back from the flak home tomorrow. Drew duty pilot tonight and am now busy doing nothing but signing for orders covering tomorrow's mission. Have written several letters.

Wednesday, September 27th

Was relieved by the new duty pilot about 0745, ate breakfast, and went right to bed, getting up for dinner and going right back again. With a fire in the stove and four blankets, I was still rather cold. Had several letters and wrote a few tonight. My enlisted men all came in tonight, and they really enjoyed their stay at the flak house. We have been alerted for tomorrow, so I am turning in early.

Thursday, September 28th
Logged 8:50 Alt. Formation (25,000)
Merseburg, Germany

Called for a 0430 briefing and sent after an oil refinery at Merseburg flying #2 of the high squadron low group with a load of 10x500 pound GP. Had a rough lead ship and the formation was hell to fly; about half the time we were in prop wash. The flak over the target was heavy from big guns but we were bombing PFF; our group made out OK. I stayed well out after bombs-away and only picked up one hole well out on our right wing.

There were lots of fighters up today and I saw them make a pass at a wing just ahead of ours, taking several down with them. I monitored the fighter frequency on VHF and heard all the talk between fighters and bombers; lots of our boys were in trouble. Had a terrible headache after the mission and went right to bed.

Friday, September 29th

Got up early this morning for an assembly but, as usual, there was a screwup and we didn't have one. Bought my rations and this week we got a can of fruit juice. Had a lecture from some economist from the Eighth Air Force on percentages of bombs in different circles in the target area, and then he chewed us out for not flying closer formation over the target.

Spent the afternoon trading electric equipment, for our heater went out yesterday and almost froze my hands. Tried to get some flak shots from the photo lab but wasn't able to. By the way, early this morning we had a crash abort and a buzz bomb went directly overhead. Everyone in our barracks ran out but I stayed in the sack, for this is as safe as anywhere else.

Saturday, September 30th
Logged 7:00 Alt. Formation (23,000)
Bielefeld, Germany

Called at 0400 and briefed to go after a sub depot at Bielefeld, Germany, but the bombing was instruments (PFF) and we hit a marshaling yard in the center of this town. Flew #3 of the lead squadron, lead group, carrying 12x500 GP. No flak, no fighters, but one hell of a lot of prop wash. I finally made a good landing for a change. This was undoubtedly a milk run, but I looked for fighters all day, and they hit the group ahead of us coming out.

Sunday, October 1st

Slept late and got up just in time to eat dinner. Had Link scheduled this afternoon but didn't get it in, for the instructors were all taking the day off. Checked out the squadron jeep and got a load of wood. Fixed a flat tire on my bike. Had ice cream for supper tonight.

Tonight was party night and as a special feature, they had a leg show. But this stunk, and the girl who did the strip tease was as flat-chested as I am. The magician was fair. Had several drinks and a few dances, then left. Went out and found a coal bucket and stole a big box full of coal. Borrowed someone's jeep from the party to haul this in and then returned the jeep. Today was the first time I have driven in the ETO.

Monday, October 2nd

A dull day with nothing scheduled. Had breakfast here in the barracks: cookies, crackers, and juice, all from home. Went down and tangled with the finance officer over a refund on the flak home deal, and it looks as if we will only get $1.25 per day when it cost us $2.00, so we lose money on this deal. Purchased a $75 money order and sent it to my wife.

Started to Norwich but couldn't get out of Attleborough, so returned to base in time for supper. Had a squadron meeting for officers tonight and they ate us out about ground school. Sat around the stove and shot the bull for several hours tonight. The day was a nice scorching one, something rare in England.

Tuesday, October 3rd

Got up for breakfast and found out I had gunnery school for the entire crew this morning at 0830. Shot skeet and broke only 18 of my 25 shots. Stopped by finance and drew our $8.75, but the enlisted men couldn't collect theirs today. Caught a ride to Norwich after dinner and left two rolls of film to be developed.

Stopped by Stone's to pick up some Scottish tartan but it still hasn't come in. Tried to bargain with him to swap some cigarettes for the goods. Had lots of mail, getting five letters from Mary Liz, and really enjoyed reading them. In one letter from my wife there was a good snapshot of her sitting in a beach chair in her bathing suit. Her suntan shows up and she looks so lovely and warm.

Wednesday, October 4th

Rained all morning so I just stayed in the sack and logged time as the rain fell on the tin roof. Took the crew to ground school at 1330 but the instructor said we were scheduled at 1500, so I told him off and said we weren't about to go out in the rain again for any class. Patched the front tire of my bicycle and I think this time it's for good. Stole a big sack of coal today so I can stay warm for another spell.

Tonight I have a big fire going, and the 1944 World Series started between the two St. Louis clubs: the Cards and the Browns. The Browns won 2 to 1. The game is coming in clear here and you can hear the crack of the bats as they rap out the old ball. A good game. Lost ten shillings in the pool tonight.

Thursday, October 5th
Logged 6:15 Alt. Formation (24,500)
Munster, Germany

Called for briefing at 0430 and went after an airfield near Munster, Germany, flying #3 lead squadron, low group, carrying 12x500 pound incendiaries. Had a Lt. Mavas from the 730th for my copilot today. No flak except at the target and no fighters. The flak was very light but accurate. Our ship picked up two holes, nothing bad. Had some new fellow leading our group and he made things rough, for we pulled 23,000 rpm all the way in. The assembly was piss poor and our group didn't assemble until after we left the field on the scheduled course. Coming back, I made a piss poor landing, bouncing three times big, but she finally settled down. Slept all afternoon and had steak for supper. The Cards won the series game tonight three to two, and I lost ten shillings more in the pool.

Friday, October 6th
Logged 1:00 Link

Got up for breakfast and had fresh eggs and hot toast for a change, which was really good. Flew the Link for one hour practicing instrument letdown, and I must say these English have a good setup for instrument procedure. Picked up a few pictures at the photography shop. Looked for another stove for our hut but couldn't find one. Packed Lt. Corcoran's clothing and took them over to the hospital, for they are sending him home before long. He lived here in the barracks and was hurt while I was at the flak home, getting broken glass in his eye. The World Series didn't come in too clear tonight, but the Browns beat the Cards 6 to 2. I lost another ten shillings in the pool.

Saturday, October 7th
Logged 8:00 Alt. Formation (25,500)
Merseburg, Germany

Called for briefing at 0330 and went after a synthetic oil refinery just southwest of Merseburg carrying 10x500 pound GP and flying #4 lead squadron, low group. Took Lt. Heine of the 728th squadron as a copilot on his second mission. No fighters and no flak except in the target area, which was inaccurate. For a change, we didn't pick up any holes. The flak over Merseburg and Leipzig looked like a cloud, and I am glad we didn't have to go through that. Fighters hit some of the groups. My neck almost broke sitting there looking through the top of the cockpit.

Returned to base and had to do an instrument letdown through a solid overcast breaking out at 1900 ft. We fly here in weather they wouldn't think of letting you off the ground in back in the States. So our pass, which was supposed to start today, has been postponed until tomorrow. We were a pissed-off bunch when they called us today. Checked out the jeep and stole two barracks' bags full of coal, so we will stay warm for a while anyhow. Went down and looked over some Halifaxes that were forced to land here when their home field socked in.

Sunday, October 8th

Had fresh eggs and toast for breakfast. Took several pictures of the Halifaxes and got some rounds of ammunition from one of their guns. Caught the three o'clock to London getting in here about 1900. Lt. McCann and I found a room in the Red Cross Club. Took Mac out to see the sights, for this is his first trip to London.

Had a few drinks before the bars closed. Stood at Piccadilly Circus and was propositioned several times by the Piccadilly Commandos, who tried to sell themselves for £5 and more. This beats all I have ever heard of before, and, to top it off, some man came by selling rubbers and was yelling, "Condoms boys, you will need them."

Monday, October 9th

Took another tour of London with Lt. McCann and really enjoyed seeing and hearing the history of such things as St. Paul's, Westminster Abbey, London Tower, etc., again. Ate at Kessler's, a French restaurant, and had a fair dinner. Went to the PX looking for some shoes, but they didn't have my size. Spent the afternoon messing around town and was somewhat amused when we priced stuff in the various shops, for they were higher than a monkey's tail.

Caught the train to Cambridge and got a room at the officer's Red Cross club. Cambridge looks good with its lights on. The Cards won the World Series by downing the Browns tonight 2 to 0. This was a really good ball game.

Tuesday, October 10th

Caught a ride from Cambridge back to camp and ate my first good meal for two days. Went to Norwich and picked up my film and ribbons and an 8th Air Force patch. Went to critique and heard an account of all the past missions, but the bad points weren't covered, as usual. Had a package of cookies from my wife today, and they are delicious.

Was in bed when the stand down came in, but the boys found a quart and woke me up and we killed this in short order, everyone getting to feeling good and shooting the bull until well into the night; about 0200. Mixed this Scotch whiskey with grapefruit juice, which made a fair drink.

Wednesday, October 11th
Logged 4:00
Practice Formation (10,000)

Called at 0600 and flew four hours practice formation flying #3 low squadron, high group, but the weather was terrible. P-47s and P-51s made simulated attacks, our gunners tracking, and we made a bombing run over a target. Landed hot and the brakes didn't hold, so we went off the end of the runway but taxied on around OK.

Had ground school this afternoon studying dinghies in case we have to ditch. It rained and was stormy all day. It's miserable outside. Broke the fire brick out of our stove, so some heat can come out and give us a little more warmth. Bought my rations. Had an excellent bull session and covered all the fields ending up on the usual subject — women.

Thursday, October 12th
Logged 8:00
Alt. Formation (25,000)
Bremen, Germany

Called for briefing at 0400 and sent after an aircraft factory near Bremen carrying 12x500 pound incendiaries flying #4 lead squadron, low group. Had Lt. Baldwin of the 730th for a copilot, and he was an excellent formation flyer.

Twelve B-17s in formation on a night mission. Photo from the Irwin Math Collection [Acc. # 2008.0151] / National Museum of the Mighty Eighth Air Force, Pooler, Georgia.

At the IP where I was all set to fly the bomb run my oxygen hose pulled out, and I passed out and didn't come to until I reached the enemy coast coming home. This was one hell of a funny feeling, going over the target and flak without knowing it. Sgt. Lundstrum plugged my oxygen in for me and brought me around. Felt OK after I came out of this. They tell me the flak was light and inaccurate. This was my 30th and DFC mission along with the tail gunner, Sgt. Claybaugh. Came in under a 1000 ft. ceiling with only two-mile visibility.

Friday, October 13th
Logged 2:00
Practice Formation (10,000)

Called at 0600 and briefed to go after a marshaling yard in Cologne, Germany, but before takeoff, this was scrubbed. (Happy day.) Took a few pictures of the crew around a B-17. Checked about our DFC, and we get them sometime next week. Was interviewed by 8th Air Force for some personnel records they are compiling. Flew lead of the low squadron, lead group, on the practice missions. My takeoff stunk as I zigged across the runway, finally getting airborne after melting the tread off our right tire. This scared hell out of me. On landing, I was cut out of the pattern three times but finally made it down. Squadron meeting of combat men, and we were all chewed out for being late to ground school. Tonight the weather is terrible, and the buzz bombs are coming over. Have seen several in flight and heard six or eight explode, one very close.

Saturday, October 14th

Got up this morning and went to two classes for ground school. One was AFC*, and Gee Box ** was the other. Looked over a de Havilland DH98 Mosquito weather ship, which landed here last night. This is a nice-looking ship, but the cockpit is awfully congested. The brakes on this ship are located on the wheel. Another two classes this afternoon in chemical warfare and flying safety. Was called to slow time an engine on a ship but talked my way out of this one. Took my leather jacket over to have a picture painted on it and also the name we have selected for our ship, "Dream Gal." Stole some coal for fuel. Early this morning, several buzz bombs went over and again this afternoon; several fell rather close.

*AFC was Automatic Flight Control.

**Gee Box was a British system to improve navigation and bombing accuracy similar to Loran.

Sunday, October 15th
Logged 6:00
Alt. Formation (23,000)
Helgoland Island, Germany

Called at 0500 and briefed to go after the installation and personnel on Helgoland carrying 30x260 pound fragmentation bombs with some new extra powerful explosive charge in them. Had a copilot and navigator from the 728th squadron. This finished Allen on 25, and this puts him and Heppner both ready for home. I flew lead ship of the high squadron, lead group.

We bombed PFF through a solid overcast, and there was very little flak. This was to clear the way for our robot bomb, which is a B-17 loaded with explosives. It is flown by radio control into the target and is supposed to follow us into this target today. Didn't pick up any flak holes for a change. This was a milk run, and I could use some more of these.

Monday, October 16th

Called for ground school this morning and went to three classes. Bought my week's rations on the way home. Went into Norwich and bought my wife two yards of Scots plaid, which cost me £4 ($16.00). Also bought her a fancy calendar as a souvenir. Had some film developed for Lt. McCann on a rush order, and I gave them a carton of cigarettes to rush the photos through. He left for home this afternoon, and I think he really hated to leave; he was sure a swell kid. Several buzz bombs came over tonight, one going directly overhead, one hit somewhere very close, for it shook the lights here in the barracks.

Tuesday, October 17th
Logged 7:00
Alt. Formation (27,000)
Cologne, Germany

Called at 0230 and briefed to go after a marshaling yard in Cologne carrying a bomb load of 34x100 pound GP and 4x500 pound incendiary clusters. The target was right in the middle of town, and I never heard of bombing a marshaling yard with such a bomb load; I think we just went after the city. The takeoff was before daylight, and I certainly had a time, for my needle and ball and gyrocompass said I was level and going straight, but the gyro horizon said I was in a 45-degree turn. Found out we had a sorry gyro made by Jack and Heintz. Shortly after this, I got vertigo, but I pulled out of it OK and broke through the fog and soup at 10,000 ft.

The flak over the target was moderate and not too accurate. We picked up only one hole. There were no fighters. But coming back from the rally point our lead navigator screwed up and drug the formation over the flak at Koblenz, and this was accurate and tracking. Four bursts went off so close under our ship the concussion bounced me off the seat each time. I was flying #4 of the lead squadron, lead group, and the flak gunners here shot my #6 man down; my crew counted six parachutes out before the ship blew up and went down in flames.

When our ship jumped, I thought sure we were hit hard, but we only picked up one more hole. Had a nice package from Sam, Jr., which was a box of Heath bars, two cartons of gum, and some razor blades, as well as a camera case, which I really need. Some more buzz bombs tonight but none of them were very close.

Wednesday, October 18th
Logged 8:00
Alt. Formation (26,000)
Kassal, Germany

Called at 0300 and briefed to go after a tank factory at Kassel carrying a load of 6x500 pound GP and 6x500 pound incendiaries. We formed at 23,000 in dense contrails, and I lost the formation but caught on as they passed the departure point on time, on course. Flak over the target was heavy but inaccurate as we bombed PFF. We had a 100 m.p.h. tailwind going over the target, so we really went.

Shot at over the battle line going into and coming from the target. Going in took only one hour, but coming out took three due to wind change. The formation was lousy all day. This one finished Sgt. Ackerman with 35, and he was one happy fellow. Got to feeling no pain later as he had several shots.

Thursday, October 19th

Slept late this morning then packed a package for my wife, which I mailed this afternoon; the main thing I sent was the Scots plaid and some pictures taken on some of our combat missions. Had another squadron meeting, and they chewed us out for not attending ground school and other things. Attempted to have some pictures made, but the weather was too bad for public relations to take the crew pictures. One of our gunners, Sgt. Hames, who quit with 25 missions, got his orders to go home today. Went out and got a leather jacket for Sgt. Ackerman. Looks like I may fly again tomorrow so I must get some sleep. Tonight more buzz bombs are coming over as some of the men are in the shelters.

Friday, October 20th

Rained like hell all morning, and it's awfully cold along with all this damp weather. Strong winds all day. Sgts. Ackerman, Claybaugh, Lundstrum and I all went over and received our DFCs. Had to go to the 96th Bomb Group, which is 45th Wing Headquarters, and some Lt. Colonel pinned them on us. I think his name was Hayes. Our pass is supposed to start tomorrow and I was highly pissed off when Operations came by and told me I was flying, but I guess I may as well take this one, and now I am hitting the sack early.

Saturday, October 21st

Slept through breakfast for today's mission was scrubbed, then showered and cleaned up to go on pass. Wrapped my DFC up and mailed it home. Sgt. Hames left for the redistribution center early this morning and didn't wake anyone, for he left at 0600. Ate dinner and started catching rides toward Boxted but the weather was poor, and I only made Ipswich. Saw the remains of a B-17, which went off the end of the runway with a full gas and bomb load and exploded, scattering parts over about ten acres. This was at the 34th group and I understand all but three of the men got out OK. Staying at the Red Cross Club here in Ipswich. Took in an English stage show tonight and it was really sorry. Turned in early.

Sunday, October 22nd

Somewhat foggy all morning and I had to walk to the railroad station, for no busses run here in Ipswich on Sunday. In the process I took the wrong street and never did see the station. Caught a ride right to the Boxted base with an English major, finding Jack Bateman without any trouble. Spent the afternoon talking with the fighter pilots, which I thoroughly enjoyed. Their P-47s look like good ships to me. Went out to a Canadian hospital tonight, for Jack wanted to see his nurse*. I spent the time talking with all the nurses and really had fun as I learned a lot about Canada's customs and habits, also lots of history. The Canadians act and think much more like the Americans than the English.

*Jack later married his Canadian nurse, Kit.

9

ON MY LAST MISSION

Monday, October 23rd

Ate breakfast with Jack this morning and started back for my home station and, with good luck, I caught a ride into Norwich reasonably fast. The weather was awful. It was very foggy with misty rain all day long. Ate dinner at the Red Cross Club in Norwich and caught the bus back to the base. Lt. Alex Bebenin, our old bombardier, whom we left in Russia as an interpreter, came back today and is now staying in our barracks. Had six or eight letters. All the boys are anxious to get through with their missions, but the group hasn't flown since my last mission.

Thursday, October 24th

Tried to shoot a round of skeet today but they wouldn't let me shoot. Picked up a picture at the photo lab. Was scheduled to fly copilot for the group C.O., Lt. Colonel Batson, as he looked over the practice formation, but some Major rolled me for this job. Got the crew together for a crew shot and Sgts. Ackerman, Claybaugh, Lundstrum, and I had our individual D.F.C. shots made. Called Ben tonight and talked with him for a few minutes. We are alerted for tomorrow, so I am turning in early to get lots of sleep.

Wednesday, October 25th
Logged 7:30
Alt. Formation (25,000)
Hamburg, Germany

Called at 0430 and briefed to go after a synthetic oil plant at Harburg, about six miles south of Hamburg, Germany. I flew lead ship high squadron, lead group, and the bomb load was 20x250 pound G.P. Today we finished our engineer, Sgt. Lundstrum, and he was one happy fellow. Our copilot, Lt. Wood, was certainly a stinker, for I let him fly three different times and he got me into hot water every time; once, we almost sat right down on the lead ship. Today two chaff ships flew ahead and rained out the chaff over the target area, and then all of our formations went in at different altitudes. These factors really confused the flak gunners, for though lots of flak came up, it was inaccurate, and we only picked up one small hole. Bombing through a 10/10 overcast, I suppose we hit the target, for a dark gray smoke came up right after bombs away. Made a crazy approach but managed to cut into the traffic pattern and land without going around. Lundstrum took some pictures with my camera, and I hope they come out OK. Am going to bed early for I am supposed to finish as copilot on tomorrow's mission, which is sure a crock.

Thursday, October 26th
Logged 7:00
Alt. Formation (26,000)
Hanover, Germany

Called at 0500 and briefed to go after a tank factory at Hanover, Germany, so I started and ended by going to the same town. I flew as copilot for a Lt. I.E. Smith but made the landing from the right seat. We flew #2 low squadron, low group, and carried 5x500 G.P. and 5x500 incendiary clusters. We didn't have any flak over the target but had a rather close call going in at the coast as they traced us with about six bursts of flak about 75 feet away. Dropped P.F.F. as there was a low 10/10 undercast, and the German fighters didn't come up. Left the formation just as we crossed the enemy coast coming home and arrived ahead of the group at 200 mph and were the first ship to land. This finished my tail gunner, Sgt. Claybaugh, as well as myself, and our bombardier for today, Capt. Frost. Took some pictures though it was rather dark. I think they will come out. All of my crew was sweating me out and waiting at our revetment* when we taxied up. It certainly felt good to be through with my tour of duty.

Bought two quarts of Black and White scotch for a cool $25 (£7) to start our finishing off party. All the officers of this barracks, and all of my enlisted men, got started on this and, after going back to the club for more scotch, we all ended up plenty drunk. We drank eight quarts of scotch and one quart of vodka. Sgts. Claybaugh and Allen passed out, but the rest of us managed to stay on our feet. We had a wonderful time shooting the bull and other things. I clipped my bombardier's tie off with a pair of scissors. Went to bed and couldn't stop the thing from turning; ended up tossing my cookies before I could ever get to sleep.

*Revetment is a parking area for aircraft surrounded by blast walls to protect surrounding aircraft.

Friday, October 27th

Woke up with my head in a condition that shouldn't be mentioned, for I was still rather high. Wrote my wife and told her I was all through, then I went down and sent her a cablegram, which I hope she will be able to dope out, for I had to use a standard phrase selection and couldn't say anything about finishing. Went to Norwich, bought some service ribbons, had another set of wings put on my battle jacket, and left three rolls of film to be developed. Looks like they will bust up our crew and send us home separately. They are trying to get Sgts. Ackerman and Allen off some time tomorrow.

Saturday, October 28th

Got up early for breakfast and ate those powdered eggs before going back to the sack. Tried to fix things up so Sgts. Ackerman and Allen could stay here until we all get ready to leave, but they will have to leave early tomorrow morning. It has been cold and wet all day, and I almost froze as I took a shower. Wrapped up some flying equipment and mailed it home to Mama. Lt. Alex Bebenin left for a week at the flak farm. Borrowed £5 from Lt. Donovan and tomorrow morning early, I am leaving for Scotland to see what the country looks like.

Sunday, October 29th

Got up rather early to see Sgts. Ackerman and Allen off as they left for Stone on the morning train. Shaved and caught a ride over to the 388th Group to see Ben on my way to Scotland. Ended up going back to my base with him to a party we were having. As usual, the boys had a fair time, and they put on a good show running after the English girls. Had to wait two hours for our ride back, but we made it back OK. Ben and I shot the breeze a few minutes then went to bed.

Monday, October 30th

Slept until noon. Found out I could catch a train from Bury St. Edmunds to Edinburgh at 1800 tonight without a change. Went down and heard Ben give some new crews a little interrogation lecture and welcome them to the group, and he laid it on heavy. Went down to his office and shot the bull all afternoon, and I found out he was about to make full colonel. He will get some Scots plaid and other stuff he wants me to take home for him. Ate an early chow, had a lunch fixed, and Ben sent me down to Bury St. Edmunds in his car. I am now on my way to Scotland.

Tuesday, October 31st

Managed to stretch out on a seat and sleep most of last night as the train rattled along. Changed trains in Edinburgh at 0700 and arrived here in Glasgow at 0810.

Found a room at the Red Cross Club and had breakfast. Took a conducted trip by bus up around Loch Lomond, and I must say it's a beautiful place with all its rolling mountains covered in timber. Came back by Loch Long and saw the English Navy test-running some torpedoes. Came back by the docks at the Glasgow port and saw many different types of warships. Saw a Sunderland flying boat land. Went to see "Cover Girl" again and after that, to a dance. They had two fair bands alternating but allowed no drinking. Had a good time.

Wednesday, November 1st

A rather heavy fog hung very low and plenty damp over the town until almost noon. Looked all over town for some Scots plaid after stopping by the Merchant Seaman's Club and talking them out of 16 clothing ration coupons. Finally found a two-yard piece relatively late in the day. Bought several other souvenirs around town, but prices are awfully high. Went back to the same dance hall and had a good evening, after which three Air Force officers and myself went to a little pub nearby. Here we saw a rather odd sight as lots of elderly Scottish women sat around drinking their scotch. After a couple of short rounds, we all returned to the Red Cross Club.

Thursday, November 2nd

Had an early breakfast and caught the 0730 train to Edinburgh. Spent the day seeing the town's historical sights, which included Edinburgh Castle and the palace where Mary, Queen of Scots, lived. Bought two yards of Scots plaid at one place and found another piece, which I must send back for later and have mailed to me.

Took in an early show tonight and caught the 2200 train back to Ely, where I must change for Attleborough. Am now sharing a train compartment with a wounded English Army private and a drunk English sailor. The sailor is very much down on his country and keeps wishing he were a Yank. The private was wounded by a mine while serving in Normandy's Canadian unit and seemed to prefer them over the English.

Friday, November 3rd

Didn't do much sleeping last night, just short naps. Arrived back at the base at noon. My orders sending me back to the zone of the interior were waiting on me, and I report to the redistribution center about November 5th. Had five nice letters from my wife, but she was somewhat worried about not hearing from me for about ten days.

Went into Norwich and picked up some developed film and some civilian ration coupons. Got the pictures of the crew and my D.F.C. shots back today. Am sleepy, so turning in early. Almost forgot that I went by the finance office and collected my pay for last month and stopped by S-2* to get a couple of photos.

*S-2 was the intelligence section.

Saturday, November 4th

Got an early start at checking out this morning and cleared everything except for the boys who weren't in their offices. Kept some of my flying clothes. Picked up my Form 5 and service records. Had to clear some places I never heard of and finally stopped by the gunnery school and picked up five boxes of skeet shells. Bought a British money order and sent for the Scots plaid I put on lay-away in Edinburgh last Thursday; have told the boys here to send it on home to me. It certainly feels good to be getting all checked out and ready to go home to my little wife just in time for Christmas. Fixed a few things up in a package to send home to my wife and will mail it tomorrow. Got my A-2 jacket back with the picture on it really looking good.

Daddy in his leather jacket emblazoned with his ship's name and pinup.

Sunday, November 5th

Rained all day, but I managed to finish checking off the base and picked up all my recommendations. Also signed out as of today. Packed some of my things and am getting ready to leave. Went out to rail transportation and picked up my ticket to Stone. Had a good scare today when I thought my clearance papers were lost but later found them OK. Called Ben, and he sent a car over for me. Lt. Kirsch, my bombardier, went over with me to see some friends. Ben had six pieces of Scots plaid and some other stuff for me to take home for him, and we had an excellent long talk together. Had a big time at the party and had just enough to drink to talk with the big wigs and feel good. Ben sent us back home at about midnight.

10

LEAVING THE 452ND

Monday, November 6th

Got up early and started packing my things, getting ready to leave the 452nd Group. Sold Alex Bebenin my bike for £8, and since I owed him a little money, he paid me the difference. Took my footlocker and A-3* down to Operations for Capt. Jack Frost is flying down and will take them along with my B-4*.

Mailed a package to my wife and left my forwarding address at the Post Office. Almost forgot, I saw Sgts. Lundstrum and Claybaugh off at 0600. Caught the 1700 train out of Attleborough to Cambridge and am now staying for the night at the Red Cross Club. Tried to call Lt. Kirby but couldn't contact him.

*A-3 and B-4 refer to luggage/duffle bags.

Tuesday, November 7th

Caught the bus from Cambridge to Rushden, and from there I caught a ride out to the 92nd Group at Podington to see Frank Farnbach, whom I found after calling all over the base. Bought my weekly rations and turned them over to him. Spent the afternoon in his barracks talking about home and things in general. We caught the liberty run into Northampton, where we ate supper, and I located a room at the Red Cross. Then we messed around town until he had to leave at eleven o'clock. Found out I had been sent to the enlisted men's Red Cross, but I slept there just the same on the hardest bed I have ever slept on. There were no springs, just a piece of plywood and thin mattress and then me.

Wednesday, November 8th

Caught the train from Northampton out to Stone, where I am now. This is only about four miles from where I was when I first came to England. Found Sgts. Ackerman and Allen and had a nice long talk with them. Will try to arrange things so we can go home on the same ship. Ran into Lt. Shedlock, a boy in my basic class who washed back a class when he told the C.O. he was afraid and didn't want to fly, but he has completed his missions. Also saw Lts. Wanderling and Shell, who were with me at Sioux City. Then I saw Capts. Frost and Dyer, who have just come down from our squadron, and I found all my baggage OK. Looks like I will be here two weeks with next to nothing to do, but I can stand that as long as I am going home.

Thursday, November 9th

Got up for breakfast and attended a lecture, which is the only thing we have scheduled here. Tried to take a physical, but they only give these at 1000 each day. Tried to get a haircut, but the barber was taking no more business. This bunch is independent as all hell. Saw some good craps and poker games going on but stayed clear of them. Found Sgts. Ackerman and Allen, and had a nice long talk with them; all enlisted men here pull guard and K.P. regularly. Found out I pick up my orders at Ft. McPherson, Atlanta, Georgia, which is out of my way. Took a good hot bath in a tub where I just laid around and soaked, and now I am turning in rather early.

Friday, November 10th

Took my physical this morning, which was nothing more than a short arm. Filled out a set of locator cards at the P.O. Listed the contents of my footlocker, and managed to get all the extras through except a pair of sunglasses. The inspection is very lax, and you can get by with almost anything. Took all pictures and film up and had S-2 censor them. Bought my rations for the week. Played blackjack all evening until midnight, and, for a long spell, I stayed about even. Then I felt lucky, took the deal, and ended up winning £65 ($260), which wasn't a bad days' work and almost put me even from the disastrous days I had upon arriving in the E.T.O.

Saturday, November 11th

Tried to get a haircut again but failed when the barber would take only twenty men for the morning. You have to get down and beg for anything around here. Bought a $100 money order and sent it to my wife. Saw where Harry Terrell registered here on September 30th. Took the powder out of a Russian 20MM shell and now have it ready to go home. Spent the afternoon reading the book "Call House Madam," and she really gives the lowdown on houses of prostitution.

Played poker for about four hours tonight, lost £10 ($40), and I have never had such piss poor luck, for I was never ahead of the game. These games are rather international, for we use both pounds and dollars.

Sunday, November 12th

Ate an early breakfast and sweated out a line for two hours to get a haircut, and, for a change, the barber did a good job. Located a good parka with the down quilting and latched onto it. Got into a craps game this afternoon and lost £30 ($120), and my luck was awful, but if I can get in a game and get lucky, I can win plenty, for there's lots of money changing hands. Spent the evening reading and then closed by writing several letters; I should write lots more, for I haven't written anyone since finishing my missions.

Monday, November 13th

Slept through breakfast, got up around 1000, and took a good hot bath, which made me feel like a million. Found out Sgts. Ackerman and Allen were shipping out, so I went over to see them off, and we talked for about an hour. Got in a ten-shilling limit poker game and lost £10 ($40) without winning a hand, so I am now almost broke and only £5 ahead, hope my luck changes before I leave this joint. Took in a show, Bob Hope in "I've Got Them Covered," which I had seen once before, then came home and read a book and wrote letters until time to go to bed. It's raining now as it has been off and on all day. Haven't seen the sun since I got here.

Tuesday, November 14th

Slept through breakfast, then went over and filed an application for a $100 partial payment, which I will get Thursday. Will take another whirl at poker, for if I can only get lucky, I can go home with plenty. As I was preparing to leave the mess hall at noon, I ran into Lt. Floyd Kirby, which was sure a surprise. He is in my barracks and will probably go home with me. We spent the afternoon and evening talking about old times, and I think I have talked him into coming down to see us while he is on furlough. While talking, I did some sewing on my blouse and short coat, which I had put off for a long time. The sun was out for a short period this afternoon for the first time since I reached this station. Managed to write six or eight letters this morning and must write more tomorrow.

Wednesday, November 15th

Slept rather late then took a good hot bath, which made me feel 100 percent better. Wrote a few letters then went to dinner, which was fair. The Canadians put on part of their USO show "Meet the Money" today, and it was really good. They had some good talent and jokes that were rather raw. Spent the night shooting the bull with some of the boys back from Sweden who were telling their tales. Apparently these boys had a big time, for they tell some interesting stories.

Thursday, November 16th

Slept through breakfast again and then stopped by finance and picked up my $100 partial payment in good old American $10 bills. Two fellows moved out of my room today and Kirby moved in, so we are again staying together. Got in a two-shilling (40 cents) limit game and lost $40 without winning a pot, so I quit for supper, then came back again, and my luck was still bad for I lost another $15, but this time I won a few pots. Looks like my luck isn't right, so I think I will lay off playing for a while. Raining now and colder than the devil. I come out on order #76 to station #15, which is Atlanta, and should be leaving here Sunday or Monday, which will put me home about right.

Friday, November 17th

Slept until almost noon, for it rained like mad all morning. Packed my A-3 bag again, filling out a list of its contents, and placed a quilted down-filled parka in the bottom, which I hope gets through, for it will be just the thing to duck hunt in. Went to the show to see Betty Grable in "Pin-Up Girl" for the second or third time, but I really enjoyed the music. This bunch certainly makes a noise in a theater, which makes the show even more interesting. Spent the night writing letters and reading a couple of magazines.

Saturday, November 18th

Slept late this morning as usual. Ran into Lts. Anderson and Nickelson, two boys who came over with us from Sioux City. They were in the 729th Squadron and went to Sweden on June 21st while on their third mission. They tell some tall tales about Sweden and have lots of good pictures to show what a good time they had while there. All the boys coming back look like they were well cared for while staying in this neutral country. Had my A-3 bag checked and put with the rest of my baggage. A glider pilot moved in the room with us tonight, and I found out he and Paul Swink* were in the same squadron.

*Swink grew up in McIntosh.

Sunday, November 19th

Had a fair breakfast and spent the morning reading the English papers and writing letters. Went to the show and saw Alice Fay in "The Girl I Left Behind." Though I had seen it before, it was good a second time. Acquired a nice-looking field jacket, for when the F/O (First Officer) moved out of our room, he left her hanging on the door. All of shipment #74 left today, and we should be cleared for shipment most any time now. Spent the night here in the room shooting the bull and reading magazines.

Monday, November 20th

Got up for breakfast and for a change we had fresh eggs: they were really good. Went over and got my rations and gave my cigarettes to Kirby. Found a deck of cards, so we played a little casino. Took in the show at 1500, "Wing and a Prayer," which passed the time though I had seen it before. Kirby was put on an air shipment and leaves at 0600 tomorrow morning; borrowed $40 off him and then we sat around and shot the bull until rather late. I then took a good hot bath and am now just before hitting the sack. Rained all day, and the sun never broke through.

Tuesday, November 21st

Intended to have breakfast with Kirby this morning, but when he woke me at 0530 I just told him goodbye instead and went back to sleep. Finally got up at 1000, took a hot bath, shaved, and then went to dinner. Played volleyball for about two hours this afternoon and really enjoyed the game; it's the first exercise I've had in months. Played poker tonight and lost $10 before I decided my luck was still bad and quit.

Wednesday, November 22nd

Slept rather late then went over to the officer's lounge, read the paper, and wrote several V-mail letters before noon. Took in the show and saw "Always in My Heart," which wasn't much, but to make things even worse, they had some stinking English short, which was awful. Had to take all our stuff out of the dressers for they took them out and returned them to the English. The Limeys will probably resell them to us through lend-lease at the price of new ones. We are really taking a licking on this job from them. Shot craps for a while tonight and won almost $40 for a change.

Thursday, November 23rd

Today was Thanksgiving and we had a swell dinner with turkey and all the trimmings, but the catch was you didn't get too much turkey. I managed to get full. Wrote a few letters home. Rumors are coming through now about our moving, but I don't expect to move before Sunday. Sat around and read, for it rained most of the day. Ran into W.W. Wilson tonight, a cadet who went through primary and basic with me, and we had quite a time talking over the old bunch. Won £2 in a poker game tonight, then quit when my luck went bad. If I can have a little more good luck, I may win back what I have lost since getting here.

Friday, November 24th

Went to the show to pass the afternoon and saw "Forever in My Mind," which I had seen before leaving the States. Tonight I shot craps for a while and lost £12, then got into a poker game and was lucky enough to win £14, so I ended up $8 ahead. Stopped by finance to see about a partial payment this morning and was rather outdone when they said we couldn't get one unless we were on shipping orders. They took our clothing closets out this afternoon, and now we have nothing left but our beds and things, which are scattered around wherever you can find room to put them. What a mess. Capt. Carmack and Lt. Crawford from the 731st

Squadron came in late this afternoon.

Saturday, November 25th

Bought and read Stars & Stripes and Yank to pass the morning. Shot craps this afternoon and again tonight and ended up winning about $20. But I could have quit once with about twice that much. Came home, took a good hot bath, and am now going to bed. Today was rather cold, and all afternoon it tried to snow and, though quite a lot fell, none stuck.

Sunday, November 26th

Slept through breakfast. When I got up, I went over to get a haircut, which I managed to get after sweating out a line for about two hours. Went to the show and saw "Buffalo Bill" for the third time and it wasn't too bad, for it passed the afternoon. Was going pubbing tonight, but this was nipped in the bud when some sergeant wouldn't give me a pass after I had cleaned up and dressed. This pissed me off no end, and I found out to get off the post, we have to sign a roster before 1300 each day. Spent the night playing solitaire and reading a magazine.

Monday, November 27th

It rained off and on all day and is still going strong. Read magazines and the paper all morning. Went to the afternoon show and saw "Standing Room Only," which I saw before leaving the States. Sat around the room here all night and shot the bull with a bunch of the fellows. Played solitaire off and on all day as if I were paying $5 for the deck and getting 50 cents for each card played, and ended up about $90 in the hole.

Tuesday, November 28th

Turned out a laundry this morning, for I was running rather short of clothing. Another shipment behind ours left this morning, making six shipments that came out after ours and are already gone; something surely has screwed up. Coming back from dinner someone called me, and it was Jack Bateman who just came in last night. Thought he was home by now but his orders were slow coming through. We sat around most of this afternoon and tonight chewing the fat. There was a heavy fog hanging over this place all day and the sun never broke through.

Wednesday, November 29th

Beautiful day out, and I took my camera around with me and snapped some pictures. Our shipment (#76E) checked personal baggage this morning, but still no word on when we leave. The show changed again, so I went over to see "Bachelor Mother" with Ginger Rogers, which was an old picture but I enjoyed seeing it again. Jack and I spent most of the day together talking about things in general.

Thursday, November 30th

Back to typical English weather today with a low overcast. Had a USO show here tonight with three women and three men, which was rather good and gave me some good laughs. Got in a payday craps game but my luck was bad and, after losing $48, I decided I had better stop. I watched one lucky duck draw about $1,500 out of the game and then quit; I have never seen anyone that lucky before. Heard from a reliable source we are leaving tomorrow afternoon sometime.

Friday, December 1st

Called to start clearing the post at 0930, but I went up and spent a couple of hours in the lounge and then made a quick trip through all the clearing points after the lines had gone. Jack and I shot the bull until 1530 at which time we ate chow and drew our rations for the trip (six cans of C-rations)*. Left the base by GI truck at 1700 and waited on the train in Hanley for better than two hours. The train finally pulled in and we packed aboard, filling every seat and now, as we roll along toward an unknown destination, everyone is trying to catch a little sleep, either sitting up or sprawled out on some stack of baggage. There are eight of us on this train going to Atlanta.

*Combat rations for when fresh food was not available.

Saturday, December 2nd

Our train pulled into the station at 0700, and we found out we are just down the river from Glasgow. After an hour delay we caught a ferry, which brought us out to the Queen Elizabeth, which is anchored in midstream. We are sleeping in triple-decker bunks on A deck, which is well up on the ship. I am in room 111A, which is rather small but has beds to sleep 21 men, though only seven are assigned to this room for the trip. From the deck I can see quite a number of ships; some ten baby aircraft carriers, several destroyers, and quite a number of smaller merchant ships. The wind is strong, and all afternoon it tried to snow. Though it didn't stick here, over on the mountains in the background there are a number of snow-capped peaks.

The small boats around the harbor were pulling alongside us all day and unloading either supplies or personnel. We have around 1000 stretcher cases aboard. Besides us, we have lots of Canadians, Australians, and quite a number of English civilians. About 1900 tonight I felt the engines start slowly turning over as we slipped out down the river under cover of the night. After slipping along rather slow and easy for about three hours, we hit open water, for the full power was put on, and now the ship is rolling and lurching, and the engines have her vibrating. Apparently, they steer the ship by increasing one or more of the props, for, at times, the whole ship vibrates. It certainly feels good to be underway, and now home doesn't seem so far away. Found out tonight the Elizabeth's total weight is 93,000 tons, and she is capable of doing over 35 knots and has never been used for civilian passage, only coming into service in 1941 after being sent to the U.S. to be finished as a result of England's war.

Sunday, December 3rd

Called at 0730 for breakfast and had fresh eggs, which were really good. Went out on deck and found the weather was rather stormy and the sea rather rough. This ship rolls rather badly, and as the waves increased, some broke well up on the bow throwing plenty of water on the decks. Saw several destroyers in the distance along the horizon. Had boat drill at 1030 as we will every morning from here on to New York. Along in the afternoon I felt rather bad, almost sick, but I think my stomach was just upset after a full supper. I feel much better now. This baby is really rocking and rolling right now, and the weather is supposed to be worse later tonight. Saw Sgts. Lundstrum and Claybaugh on deck today.

Monday, December 4th

At about 0100, the ship made a big roll. It slid me to the head of my bunk and really made a knot on my head. Sgt. Lundstrum said several fellows up in the theater fell out of bed and one broke his arm, but it's no wonder, for they are sleeping five high. Passed a large convoy today and, though they were at quite a distance, I could count 20 ships at one time. Bought several cartons of cigarettes to take home for they are only 50 cents per carton. We are carrying our life vest everywhere we go as a safety measure, and any time you leave without it, the M.P. promptly sends you back. All doors to the rooms have been removed so if the ship is hit, they won't wedge or pin anyone in a room. Got a good shampoo, which made my head feel much better.

Tuesday, December 5th

We are well south now, having held a south-southwest heading most of the time, and the weather is nice and warm as the sun shone all day. As an evasive movement, the ship changes course every five to ten minutes, so we are nearly zigzagging home. Took a fresh water bath but had to carry the water to the tub in a bucket. We have a bath but both the commode and tub have only saltwater and the basins haven't been connected yet. This is a necessity for sometimes they have 18,000 troops aboard and fresh water runs rather scarce. Spent the day out on the deck reading a book, "Tidewater." Sgts Lundstrum and Claybaugh came by tonight and we spent several hours talking about things in general.

Wednesday, December 6th

This morning the weather was nice and warm, and right after life boat drill Sgt. Claybaugh and I went up and talked with Sgt. Lundstrum who was on gun guard duty. Played blackjack most of the afternoon and lost about $15. Tonight I read the rest of "Tidewater." The sea is getting rough again now and the Elizabeth is rolling rather badly, but being 1034 feet long there is very little movement at the longitudinal axis. We set our watches back one hour at midnight as we have been doing every night since we put out to sea. This Atlantic Ocean is a big pond and I never dreamed there was so much water.

Thursday, December 7th

Three years ago today all this mess started, and late this afternoon we saw our first signs of being near the States when a PBX5 on patrol duty circled the ship. We were in a rain squall all morning but the afternoon was nice and sunny with a strong wind. Got in a poker game tonight and lost my shirt as I sank my last $35. To get some cash tomorrow I will have to write a check. Maybe someday I will get in one of these games and have a little luck. It certainly feels good to be getting back near home, and tomorrow night I will call my wife. Tipped our table waiters $2 each. The ship is rather smooth now and we could make port about noon tomorrow.

EPILOGUE

Blowout at Poltava

by

Richard R. Muller

(Originally published in the February 2015 issue of World War II)

The afternoon light was bright over eastern Ukraine as the sky filled with giant silver planes. Back in friendly airspace after the black clouds of German flak fields, 74 B-17 Flying Fortresses with U.S. Eighth Air Force markings arrayed in crisp formation above the airfield at Poltava before maneuvering to land one by one, then taxiing into place on Marston matting laid over grass.

By the time the crewmen of the final Fort to set down on that cramped expanse of Soviet soil had cut their plane's engines, 45 minutes had passed.

The Forts, with their oversize "Tokyo" fuel tanks nearly dry, were not wearing the customary coat of olive drab. To wow the Russians, the Eighth had sent its newest G-model aircraft, their aluminum skin bare, as it had been since early 1944, to reduce weight and drag. From the big deadly birds streamed nearly 800 tired Americans clad in bulky flying gear — but carrying the natty Class A uniforms they would soon don. In their pockets, the airmen had maps of the U.S.S.R. and flashcards translating English sentences into Russian and vice versa. As the fliers approached a committee waiting on the field to greet them, their superiors' warnings replayed in their heads to not do anything that might make the United States "look cheap, shabby, or uncouth."

To mark the occasion, base commander Major General Alexander R. Perminov and his people had arranged a ceremony. By coincidence, the date, June 21, 1944, was the eve of an anniversary Russians held in infamy: on June 22, 1941, their and the Americans' now-shared enemy, Germany, had unleashed Operation Barbarossa — a back-stabbing surprise invasion of the Soviet Union that Hitler planned even as Germany was reaping the benefits of a mutual nonaggression pact with the Soviets. The festivities began amid fumbles. The Americans, now in their Class A getups — dark olive drab wool blouse, tan trousers, brass collar insignia — had trouble interpreting Soviet insignia.

Translation problems arose. Asked to make remarks, American commander Colonel Archie J. Old, Jr. stood at the microphone. As he began speaking, a female Red Army soldier came forward and pressed a bouquet on him. Flustered, the no-nonsense colonel proceeded to "smell the flowers and talk into the microphone, then smell the microphone and talk into the flowers," the 3rd Bombardment Division signals officer recalled.

Once the talking was done, Soviet minders whisked their famished guests to a mess hall. As he awaited his meal First Lt. Stanley Bonda chuckled to hear a Ukrainian woman serving dinner say, "Damn C-rations again." But the young pilot also could see the awe on his new comrades' faces as Russians realized they were in the company of men who had laid waste to the Hitlerite capital. "The more times you bombed Berlin, the greater a hero you were in their eyes," another pilot said. Afterward, the Americans retired to tents, with slit trenches nearby; Ukraine was a war zone after all. On that cool June evening, similar scenes played out at two other Ukrainian airfields for the rest of the American contingent.

The sunset, just before 9:00 p.m., ended a long journey that, all in all, had gone well — except that unknown to the Americans and Russians, the Germans knew precisely where it had ended.

Early that morning, an enormous American force had flown from England to drop nearly 2,500 tons of high explosives and incendiaries in what first appeared to German defenders to be a standard daylight terror raid. Instead of flying back to England, however, a fraction of the American bomber crews had borne on eastward with P-51 Mustangs providing cover to land at three Soviet air bases in Ukraine: Poltava, Mirgorod (50 miles northwest) and Pyriatyn, another 50 miles beyond Mirgorod. One of many western attempts, including the recent invasion at Normandy, meant to encourage the U.S.S.R. to fight on, the effort was known as Operation Frantic. And save for sharp German eyes, a wayward Mustang, and what remained of the Luftwaffe bomber fleet in the East, Frantic might have succeeded.

In late 1943, the Allies' lack of a fighter able to fly to and from Berlin protecting heavy bombers had left their daylight bombing offensive in jeopardy. American air planners sought to open a second air front from the East; heavy bombers flying out of Soviet bases would threaten the Reich from an unexpected quarter. But the Soviets, mistrustful of their allies and skeptical of the value of long-range bombardment, dragged their feet. The Americans persisted; not only were there diplomatic benefits to be had, but the Soviets maintained air bases in their far east. Access to those facilities and landing fields would permit American bombers to pummel Japan from mainland Asia rather than from the Pacific alone.

In February 1944, Premier Joseph Stalin finally agreed to welcome American crews and planes to Poltava, Mirgorod, and Pyriatyn. The U.S. Army Air Forces began planning missions that would fly out of England and Italy, hit Axis targets, and continue to Soviet territory to land and refuel before bombing more objectives on the turnaround. In addition, hundreds of American personnel deployed to the Soviet bases to work with Soviet comrades on the "shuttle" project, which eventually would include American planes based on Soviet soil.

The first shuttle of Eighth Air Force planes would take place as part of a regular daylight raid. Cleared for June 21, 1944, that mission would send more than 1200 B-17s and B-24s from England to Germany, with nearly 1300 P-51 Mustangs, P-38 Lightnings, and P-47 Thunderbolts riding shotgun to cover the enormous bomber formation in relays. After the 775-mile flight from England to the targets, the six shuttle groups — some 140 B-17s guarded by Mustangs with wing tanks — would fly on to Ukraine, a 12-hour, 1,670-mile journey. Air force planners had structured the Berlin area strikes to draw German defenders away from the shuttle bombers, which hailed from the 3rd Bombardment Division's 45th and 13th combat wings. Once those crews attacked a synthetic oil plant at Ruhland, 75 miles south of Berlin, they would bear southeast for Ukraine.

As fliers across southeast England woke for breakfast early that day, shuttle crews' excitement outran the usual raid day tension. At Deopham Green in Norfolk, the intelligence briefer addressing the 452nd Bomb Group (Daddy's group) was in rare form, pilot Stanley Bonda recalled, riffing about, "The history our group was going to make, pioneers, and blazing a trail."

"My God!" another pilot exclaimed as he studied the map and reality sank in. "That's a long way from home!"

The 452nd, which was sending 26 heavy bombers to Ukraine and expecting 19 to return to England, would be taking off at 5:45 a.m.; if the schedule held, the last plane on the shuttle flight would land in Ukraine at 4:40 p.m.

The Berlin raid was not a sham, but a maximum effort — every B-17 group in the Eighth was in. Formations converged to deploy against more than 15 targets amid a storm of flak. German fighter forces, though depleted, were still capable of seizing an advantage. One presented itself when a portion of the American fighter contingent missed its rendezvous, leaving part of the bomber stream protected poorly. The Luftwaffe threw a task force of twin-engine fighters into the gap, and six bombers fell. Flak, as usual, was intense and accurate. The Berlin raiders lost 77 bombers with 502 damaged.

"The bombing itself was below the usual Eighth Air Force standards," the Eighth's command declared later.

Things went better for the shuttle force once its bombers reached Ruhland. After they unloaded and headed southeast, production at the hydrogenation plant did not resume until many weeks later. The mustang fighters of the 4th and 352nd Fighter Groups would fly cover for the bombers until the formation was safely inside Soviet airspace, then peel off to set down at Pyriatyn. The 13th Combat Wing would land at Mirgorod. The 45th would fly the furthest, to Poltava.

For all their attention to detail, the shuttle mission's planners had ignored one significant variable: the Luftwaffe in the East. The oversight was understandable. Since mid-1943, the Germans had pulled many fighter wings from the Eastern Front to contest the American bomber offensive. In addition, Germany's eastern bomber arm was a shadow of the winged armada that had kicked off Barbarossa in June 1941. Years of tenuous supply lines and costly army support operations against the stubborn, skilled Soviet foe had taken a toll on planes and personnel. Moreover, Allied intelligence reported that Luftwaffe bomber units in the East were mostly still deploying the Heinkel He 111, a twin-engine middleweight that had first been flown in anger during the Spanish Civil War.

By autumn 1943, however, things had brightened for a few hundred German crews in the East. Soviet factories in the Moscow-Upper Volga region were pouring out materiel. To interrupt that supply chain, the Luftwaffe planned strikes by night against the power plants that kept those factories humming. Five full bomber wings — nearly 350 He 111s — were jerked from the attritting grind of support flying to become Fliegerkorps IV, commanded by General Rudolf Meister. The new unit trained in long-range navigation and target location and studied pathfinder and target illuminator tactics. Their venerable Heinkels had been updated with the latest avionics, engines, and armament, and crews loved them.

By spring 1944 Fliegerkorps IV, based around Brest Litovsk, was poised to strike — but a winter of Soviet advances and German withdrawals had left key northern bases unusable, placing some targets beyond the Heinkel's 600-mile operating radius. However, the Eastern Front had no shortage of objectives — notably Soviet rail lines that had gotten noticeably busier in northern Ukraine. Soviet night air defenses were weak, and Fliegerkorps IV flew dozens of barely opposed night raids against trains, yards, and rails between March and June 1944, working from reconnaissance photos taken by Junkers Ju 188 crews. Although poorly trained and unskilled replacement personnel had most other Luftwaffe units doing hackwork, these Heinkel crews were getting better.

As intended, the Berlin raid had indeed kept Luftwaffe Command busy. Nonetheless, when the shuttle group continued east after hitting Ruhland, someone noticed. German planes tailed the Americans; Colonel Old saw one dive into a cloud to escape the Mustangs. As the Allied aircraft departed German-held territory, the spotters turned back. Obviously, the Americans were bound for Russia, but where — and why? When a shot-up Mustang crash-landed in Poland, the Luftwaffe got its answers. Aboard the nearly intact P-51, searchers found papers that not only revealed the shuttle flight plan but also named the shuttle bases. Luftwaffe Command relayed Meister's information with orders to swap that night's railroad strikes for a rare opportunity.

Meister sent two Ju 188s to photograph the fields at Poltava, Mirgorod, and Pyriatyn. The high-flying reconnaissance aircraft circled the shuttle bases, easily dodging ineffectual Soviet flak. Pictures showed B-17s at

Poltava parked in neat rows — "a peacetime lineup," as an astonished Meister put it.

Planes were arranged wing to wing at Poltava for the most straightforward reason: lack of space. The base had only so much Marston matting — and no blast revetment. At Mirgorod and Pyriatyn, as enemy reconnaissance crews were photographing other Forts and their P-51 escorts, American personnel spotted the Junkers. Mustang pilots asked permission to pursue. The Soviets said they would handle it. By the time Russian pilots scrambled, however, the Ju 188s were gone. At Mirgorod, the 13th Combat Wing crews refueled in record time and took off to protect bases at Kharkiv and elsewhere. The 74 bombers at Poltava stayed put — the crews were tired, the Russians seemed unconcerned by the enemy reconnaissance aircraft, and Poltava seemed a safe distance from any German bases.

Meister and cohorts applied themselves to the evening's duties. Three bomber wings, each with three groups of 20 Heinkels fueled and loaded ordnance. Two wings — some 120 bombers — would attack Poltava, preceded by one group of 20 Pathfinders. Other Pathfinders would mark Mirgorod for the third wing. Pyriatyn was not a target; no bombers had landed there, but the shuttle force's Mustangs, a formidable fighter complement, had.

The German Pathfinders were based near enough to their objectives to operate from the home base; many of the bomber groups had to stage to airfields closer to the front line. At the last minute, one wing of Heinkels, finding itself short of time to refuel and arm, dropped out of the attack reducing the Poltava strike force by half and carrying high explosives fused to explode on impact along with incendiary, fragmentation, and SD-2 antipersonnel bombs — the last nicknamed "butterflies" because when open, their casings resembled the insect's wings. Heinkels took off from fields around Minsk in fading light. They arrayed in a "bomber stream," a tactic borrowed from the R.A.F. in which loose groups of six to eight planes flew in close succession so pilots could swamp enemy defenses at minimal risk of collision.

Ahead flew the Pathfinders, steering by electronic navigation to the Soviet bases where they dropped marker flares. However, the Mirgorod Pathfinders fell behind after three and a quarter hours in the air and unknowingly overflew their target. Spotting markers that other Pathfinders had dropped at Poltava, the latecomers assumed they were over Mirgorod and dropped flares as well. The leader of the German bomber wing detailed to attack Mirgorod recognized the error and seized the moment. The entire German bombing assault fell on Poltava.

An alert there had jerked the base's several thousand Americans and Soviets out of their cots, but not until the flares began falling and the antiaircraft guns opened up did urgency sweep the field. Men pelted to the slit trenches, some tripping over tent ropes. American crews watched transfixed as a single marker flare without a parachute dropped onto the center of the main runway.

A line of parachute flares stretched east to west, making the silver B-17s gleam. "I can still see the brilliant white flares parachuted by the German bombers swinging back and forth, lighting up the night like it was high noon," Technical Sgt. John Chopelas, a radio operator with the 452nd, recalled. The Germans continued dropping flares until the light from burning bombers made them unnecessary.

Attacking in several waves, Heinkel crews let loose some 15 tons of high explosives, including five 2200 pound blockbusters that opened huge craters among the densely packed aircraft. Another 78 tons of fragmentation bombs shredded the B-17s aluminum skin, and 17 tons of incendiaries reduced the big bombers to flaming wrecks. In all, some 6500 individual projectiles fell on the field.

One German group commander noted that the most experienced crews led the attack, carefully lining up on targets and making several approaches before releasing payloads. Crews were supposed to spend only 20 minutes over the target, but pickings were so rich many lingered twice that long.

"They seemed to be flying in a regular traffic pattern," an American said. "They would make their run, get back into the traffic pattern again, all very casually."

Below, all was chaos. John Chopelas, who had flown many a bombing run, never forgot the helpless sensation of being on the receiving end of what he had been dishing out. Soviet personnel rushed to the blazing aircraft trying courageously, but futilely to fight fires with shovels and buckets of sand, heedless of the frag bombs carpeting the scene. Poltava's 450,000-gallon aviation fuel dump caught fire, and one of the bomb dumps produced in an American's words, "very pretty fireworks."

To the Americans, Soviet defenses seemed totally ineffective. Searchlights would "cone" a German bomber pinioning the plane with beams — but gunners, lacking radar or coordination with light crews, let pilot after pilot escape. The American combat wing's lead navigator recalled how a German flier was able to "cross the field and methodically drop his bombs with A.A. fire bursting consistently 50 yards behind his tail". Poltava base commander Perminov summoned Soviet night fighters, but none appeared; those pilots did not report to him. After the He 111s had spent about 90 minutes raking the airfield, the German Pathfinder commander gave the signal: "Closing time!"

With nary a loss, the raiders departed, leaving in their wake a conflagration that a reconnaissance crew documented, dropping flash bombs to get the image on film. "The whole aerodrome was ringed with burning and exploding aircraft," said Wing Commander Hughes, an R.A.F. liaison officer assigned to the base. Only two Americans died, but that night dozens of Russians lost their lives trying to douse flaming B-17s. Others were wounded, and thousands of antipersonnel bombs littered the area, so the defenders decided to leave the desolation until daylight.

The next morning American anger at the poor Soviet showing abated as Russian personnel worked heroically to clear the field, an effort that took days and cost additional Soviet dead and injured. Radioman John Chopelas recalled Russians methodically picking up ordnance along a path to the ruined and damaged aircraft, grinning while holding up unexploded butterfly bombs for the Americans to "admire."

Of 74 B-17s at Poltava, 47 were total losses, as were two C-47 transports and an F-5 reconnaissance plane. Many B-17s burned to cinders except for the tails. Another 26 were damaged. Counting losses from the June 21 raid on Germany, this was the Eighth Air Force's costliest single operation of the war.

The Soviets lost several planes, including Perminov's. The attack had been uncannily precise; 98 percent of German ordnance hit the target area. Four days later, the 13th Combat Wing and the remnants of the 45th left Ukraine to strike a Hungarian rail yard en route to Italy and the U.K. Most 45th crews, planes in ruins, stayed a week at Poltava before riding American transports to

England via the Middle East.

The U.S. Army Air Forces at first slammed the lid — "As if the Germans didn't know what happened!" Chopelas said — but censorship yielded to writing off the attack as an expensive lesson. By July, reporters were being briefed. An official periodical ran a spread of the mess declaring that the Luftwaffe "CAN STILL TOSS A WICKED PUNCH."

The 3rd Bombardment Division took comfort in industrial production. Poltava was "a Russian graveyard," yes, but the losses were only two days of B-17 plant output, a report stated. More than a year later, a report on the P-51's lost papers warned airmen not to carry sensitive documents into combat. The Poltava affair came to be seen as a stew of inefficiency, confusion, American hubris, and German skill. The Soviets took the raid in stride; the Luftwaffe always merited respect — how odd that the Americans had been caught so flat-footed.

For the Luftwaffe pilots, Poltava briefly harkened to Barbarossa's early days. The night of June 22, the invasion anniversary, Fliegerkorps IV returned to Mirgorod and Pyriatyn. They missed Pyriatyn and hit Mirgorod hard, but the American bombers were still elsewhere. That day the tremendous Soviet offensive, Operation Bagration, began. Meister's unit returned to providing ground support, but within weeks its Heinkels were out of gas. By September, Fliegerkorps IV no longer existed.

Disaster did not end the shuttle experiment, but the handwriting was on the wall. Missions continued into September 1944. The last one was an attempt on September 18 to support the Polish underground's doomed uprising in Warsaw. The next day the B-17s left

Poltava for good. As fall cooled into winter, only a skeleton crew of Americans remained in Ukraine, closing up shop at the Operation Frantic bases.

MISSIONS

The Bombing Missions of Lt. Edwin "Snake" Walkup

Daddy was a B-17 pilot or copilot on each of these 35 missions. He was assigned to 731st Squadron 452nd Bomb Group of the 3rd Wing stationed at Attleborough, a small town near Norwich. The name of the airfield was Royal Air Force Deopham Green. During the war the 452nd Group flew 250 missions from this airfield and lost 110 of its bombers.

Date	Destination	Target
6/18/1944	Hanover, Germany	Oil refinery
6/19	Bordeaux, France	Airfield

6/20	Magdeburg, Germany	Synthetic rubber plant
6/21	Ruhland, Germany	Synthetic rubber plant and on to Poltava
7/6	Abbeville, France	Fuel dump
7/7	Leipzig, Germany	Synthetic oil plant
7/12	Munich, Germany	
7/13	Munich, Germany	
7/14	Aurillac, France	Drop supplies for the French underground
7/18	Kiel, Germany	E-boatyard
7/19	Schweinfurt, Germany	Ball bearing plant
7/21	Regensburg, Germany	Assembly plant
7/24	Saint-Lô, France	Support for ground troops
7/25	Saint-Lô, France	Support for ground troops
7/28	Leipzig, Germany	Synthetic oil plant
8/4	Bremen, Germany	Oil works
8/5	Magdeburg, Germany	Iron factory
8/7	Reims, France	Railroad bridge
8/9	Nuremberg, Germany	Diesel engine plant
8/13	Nantes, France	Support for the Canadians in Normandy
8/25	Stetten, Germany	Oil refinery
9/1	Mainz, Germany	Truck factory
9/8	Mainz, Germany	Tank factory and supply depot
9/10	Nuremberg, Germany	Tank factory
9/13	Bielefeld, Germany	Marshalling yard
9/28	Merseburg, Germany	Oil refinery

9/30	Bielefeld, Germany	Sub depot
10/5	Munster, Germany	Airfield
10/7	Merseburg, Germany	Synthetic oil refinery
10/12	Bremen, Germany	Aircraft factory
10/15	Helgoland Island	Installation and personnel
10/17	Cologne, Germany	Marshalling yard
10/18	Kassel, Germany	Tank factory
10/25	Hamburg, Germany	Synthetic oil plant
10/26	Hanover, Germany	Tank factory

For completing 35 missions, Daddy received the Distinguished Flying Cross (left). For meritorious achievement while participating in aerial flight, he received the Air Medal (right).

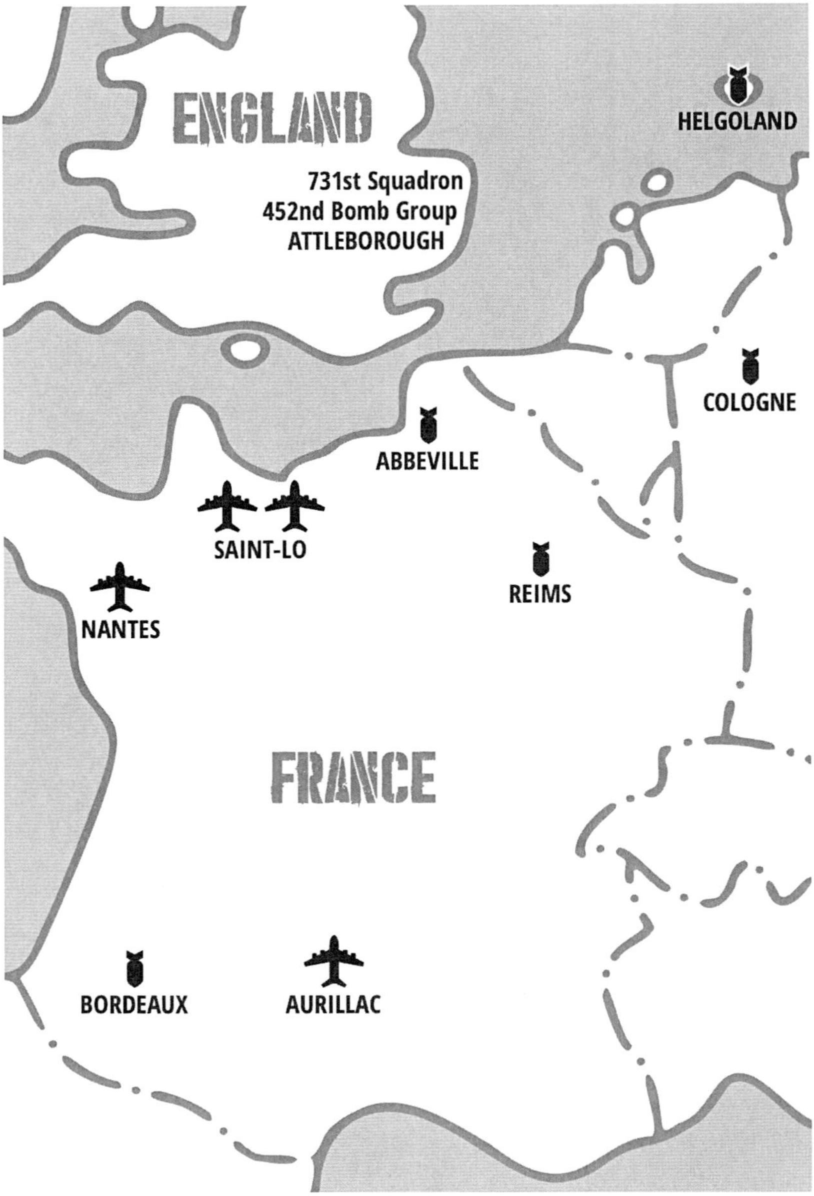

Daddy's 35 missions for the 731st Squadron 452nd Bomb Group of the 3rd Wing. Each bomb icon represents a bombing mission on that location. Each B-17 icon represents a support mission to that location.

KIEL
HAMBURG
GERMANY
BREMEN
HANOVER
MAGDEBURG
BIELEFELD
MERSEBERG
MUNSTER
RUHLAND
KASSEL
LEIPZIG
MAINZ
SCHWEINFURT
NUREMBERG
STETTEN
REGENSBURG
MUNICH

CAREER

The Armed Forces Career of Lt. Edwin "Snake" Walkup

Daddy graduated from Stetson University in DeLand, Florida with a BA degree in Business Administration. When he was drafted into the Army he was sent to Ft. Sill, Oklahoma to work in geodetic computation. One day, while marching in the hot sun, he looked up and saw airplanes overhead. He immediately applied for the U.S. Army Air Corps Cadet training program and was accepted.

His Primary Flight training was at Thunderbird Field near Phoenix, Arizona; Basic Flight training was at Pecos, Texas; and advanced training was at Ft. Sumner near Clovis, New Mexio. He graduated as a 2nd Lt. Army Air pilot assigned to B-17 bombers and got his wings there on July 27, 1943. He got a pass to go home and was married to his long-time girlfriend, Mary Elizabeth Harrison, in the Presbyterian church in McIntosh on August 1, 1943. Accompanied by Mother he was assigned for short periods of time at five different bases in Washington, Oregon, Nebraska and Iowa until he departed from Kearney, Nebraska for England on May 20, 1944.

After he came home from the war he was in the Army Air / U.S. Air Force for an additional eleven years, during which time he was promoted to Captain. In an interview years later he said that he and Mother were moved every few months to various bases in Florida. While he was in Berlin working during the Berlin Airlift helping to distribute supplies our planes delivered to Berlin, Mother and I lived with Daddy's parents in McIntosh. I remember that we lived for a while in Tampa at MacDill Air Force Base. I don't know what his job was during that time as his records were destroyed in a fire. I do know that he was on the MacDill Flight Service Skeet Shoot team, champions of the Eastern Division and entrants in the national flying service finals. He also said in that interview that he was offered another promotion if he would fly bombers in Korea and he said, "No, thank you." Daddy retired as a Captain.

TRIBUTE

"Remembering is the Tribute"
The Passing of Lt. Edwin "Snake" Walkup
Published in The Gainesville Sun, July 5, 1997

Being remembered for the good we've done is not a bad legacy. When those recollections come from those who knew us best, it has to be the ice cream on the cake of life.

And that's the fate of Edwin F. "Snake" Walkup, although many of those who knew him may not have known his real name was Edwin. When he'd thrust out a hand at introductions he was Snake, which was really an abbreviation of his full nickname.

His widow Mary Elizabeth explains that when Edwin was growing up in McIntosh, Paul Boulware was the official bestower of nicknames in the community. As a gangly 14 year-old who was a little too tall and skinny, Boulware noted that the boy reminded him of this dragonfly-like bug known locally as a Snake Doctor. The nickname lasted a lifetime.

Snake died last November and he's remembered for more than an unusual nickname. This week at the McIntosh Community House his family and friends dedicated a small memorial in his honor. "Edwin F. "Snake" Walkup, 1915-1996, Lions Charter Member 45 years, WWII 35 combat missions, Town Council 27 years" are the accomplishments carved on the small piece of granite next to a taller slab honoring local veterans.

But a few lines on stone can never tell the whole story. Not mentioned is a lumber business he ran in Archer, or his football skills back when Stetson University was playing football. Snake's cousin Howard Walkup says it was the Lions Club that was behind the memorial. Like his cousin, Howard served on the town council. "Too long" was how he measured his eight-year stint. When asked to rate which was tougher, Snake's 27 years in small-town politics or 35 combat missions over Germany, Howard laughed. "I don't know, that's a toss-up," he said.

Ned Cake found Snake's combat diary so interesting that he summarized it last fall, as Snake was trying to fight back from a stroke and a heart attack. He didn't admit to the heart attack however, claiming until his death that it was just a case of indigestion caused by an overdose of Chinese food.

The stories in Snake's journal told of harrowing bombing raids and landings in a B-17 peppered with flak. Cake says Snake's account of a raid on Ruhland, Germany, had details not found in other accounts. Specifically, how the raiding group landed in Russia to re-arm and refuel and then watched as more than 50 planes were destroyed or badly damaged in a German raid. Lewis Sherouse, a retired U.S. Air Force colonel living in McIntosh, had also heard those stories. "He was a gifted storyteller," Sherouse said. "He could put you in the airplane with him."

But it's not the wartime heroics that endeared Snake to Cake, it was the years of morning meetings at the one-acre garden plot Cake kept with a friend. Snake's gardening specialties were cucumbers and tomatoes but the conversation wasn't limited to horticulture. Friendship grew in that sandy soil as well. "We got to be good buddies with him visiting us," Cake said.

So, while the memorial mentions accomplishments,

it's really friendship that's being remembered. Snake's son, Edwin Jr., says he doesn't recall his dad having a harsh word to say about anyone. But what would Snake have to say about the fuss? Mary Elizabeth believes, "He'd be greatly pleased, but he'd have said to the Lions, 'You shouldn't have spent the money.'"

ABOARD THE B-17

Details on the Flying Fortress

- Four engines
- Length of 74 feet
- Wingspan of 103 feet
- Could fly at 287 mph
- Range of 2,000 miles
- Ten-man crew: pilot, copilot, navigator, bombardier/nose gunner, flight engineer/top turret gunner, radio operator, tail gunner, two waist gunners, ball turret gunner.
- The B-17s would drop more bombs than any other aircraft in the U.S. Arsenal during WWII.
- The name "Flying Fortress" came about because the aircraft was so heavily armed. Six of the ten-man crew were aerial gunners firing twin .50 caliber machine guns.
- Daddy was with the 452nd Bomb Group stationed at Deopham Green. They flew 250 missions and lost 110 planes. In total, there were 10,631 missions flown by the

8th Air Force with a loss of 4,754 aircraft.

- A total of 12,731 B-17s were produced with most of them being used in the ETO. About 50 are left in the world and only 12 are still airworthy.
- The 8th Air Force took more casualties in WWII than the Marine Corps and the Navy combined.
- There were about 200,000 combat crewmen in the 8th. Of those 26,000 were killed or MIA and 28,000 were taken prisoner. Half of the U.S. Army Air Forces' casualties in WWII were suffered by the 8th Air Force.
- The average age of a crew member was 20.
- According to "Fast Track to Manhood" by Thomas P. Griffin, "Flying four-engine aircraft would give a pilot the prestige in flying circles that a PhD gives a staff member in academic circles."
- General Carl Spaatz, the American air commander in Europe, said, "Without the B-17, we may have lost the war."

The B-17 Crew Positions

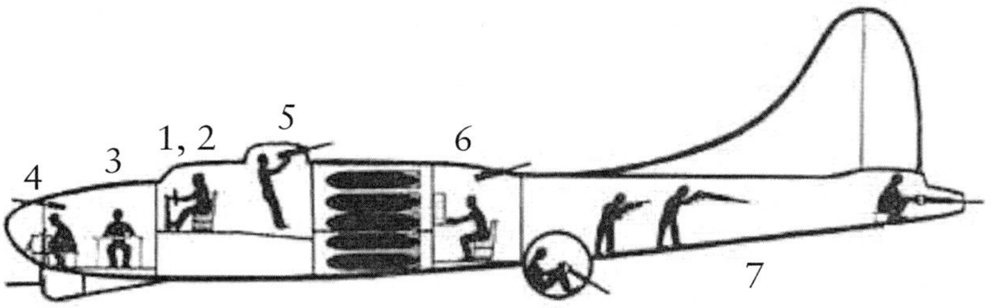

Crew assignments for each of the B-17 positions were:

1. Pilot – the pilot is the commander of the plane and its crew on a 24-hour basis. He is the one who brings them together as a team and builds the morale of the whole. He must command respect, confidence and trust in the entire crew. The pilot is the one who takes off and lands.
2. Copilot – second in rank to the pilot and generally flies the plane between take-off and landing. (There was also auto-pilot, which could be used sometimes but usually not when flying in formation.) Copilot must be ready at any moment to take over for the pilot.
3. Navigator – the one who must know at all times the exact position of the plane. In getting close to the target, the navigator must know the position of the plane within a quarter-mile of the target. He must also be able to use the plane's .50 caliber machine gun.
4. Bombardier – the one who sights in on the target and decides when to release the bombs; however, if they are flying PFF, the lead ship makes that decision and the Bombardier simply presses the release button to drop the bombs when the PFF ship does. The Bombardier actually takes over the airplane for the run on the target, and at that time he becomes the commander of the plane. He also must be ready to use the two machine guns at the nose of the plane.
5. Engineer/top turret gunner – in charge of keeping the plane running smoothly, as well as the radio and all of the guns. He is supposed to know more about the plane than any other crew member. He also has to be a good gunner.
6. Radio operator – in charge of communications with the other planes in the bombing group and must be ready to help with the machine guns.
7. Gunners – the four aft gunners have to know the firing ranges of their .50 caliber machine guns and, as the waist gunners are firing out of open windows, they must keep their guns from freezing up.

The B-17 in Literature and Film

Here are some interesting passages about the B-17, beginning with two from "The Man Who Flew the Memphis Belle" by Col. Robert Morgan.

Before each takeoff we would get into that huddle of ours on the tarmac beside the airplane and put our arms onto one another's shoulders and recite a pledge. If only one airplane comes back today, it's going to be us, guys. *I don't know if it did us a bit of good or not, but it kept us from that mindset of* maybe this is my last mission, maybe I'm going to get it today. *It reminded us that we were no longer ten individuals but one interdependent unit. Each of us is protecting all of us. Each job is to help the other jobs. The gunners are protecting Jim Verinis and me as we fly the plane. We're helping them by keeping it steady. The navigator is our eyes. The radio operator is our ears. And every man on board is ready to switch jobs, grab a machine gun, grab the mic, grab the stick, in a moment of crisis.*[1]

It doesn't matter if it's the first time or the twenty-fourth, which this one was. What matters is that you're down out of the sky. Your wheels are on the tarmac. You've brought your crew back safe.

You take a big deep breath. You feel the sweat that has soaked through your long underwear and your tunic and lubricated the fleece lining of your leather jacket, plastering it to your shoulders and back and chest. Maybe you clench and unclench your hands a few times. Man, it feels good to have turned loose of that yoke. You've held on to it for maybe six, eight hours, knuckles white, keeping thirty-two tons of bomber steady at 25,000 feet, your hands wrapped around that shuddering yoke, your feet tensing against the rudder pedals. Your margin of error is down to feet, from the wingtips of the bomber on your left and the bomber on your right. All this with deadly antiaircraft fire and cannon and machine gun rounds from hostile fighter planes and the smoke and debris from your comrades' planes around you cracking your universe apart just as the trip started to get dull.

Now maybe you start to slide your feet off those pedals, let your legs go limp a little, and here comes the pain from those releasing muscles, shooting all the way up your calves and thighs clear to your hips. Feels good, that pain. It means you're still alive.[2]

From "The Mighty Eighth" by Gerald Astor.

I could not believe the mass of instruments in the cockpit of the B-17... 146 switches, gauges, indicators and warning lights. Use was made of a "checklist" card wherein the copilot called off checkpoints while the pilot had to reply that the item was in order. This included fifty-seven checks before takeoff, during flight, and after landing. There were thirty-eight checks for "end of mission."[3]

"The B-17," notes Villani, "had no power assists. All the control surfaces were operated by cables connected directly to the pilot's hand controls. Maneuvering meant applying force to a control surface and waiting for the plane to react. This little inertia [during the brief interval between manipulating a control and effect] was what made it tough. The pilot, for example, was always inching the throttles back and forth in order to hold position and not crash into somebody."[4]

From "Citizens of London" by Lynne Olson.

> *It didn't take long for the Eighth's pilots and crews to realize that, in coming to Britain, they had been handed one of the most dangerous jobs of the war. U.S. Army Air Forces casualties, particularly in the Eighth, were astronomically greater than those in either of the two other military services. According to conventional wisdom, a crew member's chances of completing his normal tour of duty – twenty-five missions – were less than one in four. In its first ten months of operation, the Eighth lost 188 heavy bombers and some 1,900 crewmen; those numbers would skyrocket over the next year-and-a-half. By the end of the conflict, the U.S. air operation in Europe would suffer more fatalities – 26,000 – than the entire Marine Corps in its protracted bloody campaigns in the Pacific. "To fly in the Eighth Air Force in those days," recalled Harrison Salisbury, "was to hold a ticket to a funeral. Your own."*[5]

For actual footage of B-17s in action, check out *WWII in HD*, which is narrated by Gary Sinise. Episode 2 *Hard Way Back*[6] and Episode 4 *Battle Stations*[7] have great action shots, especially Episode 4.

---

[1] Morgan, "The Man Who Flew the Memphis Belle," page 135

[2] Morgan, "The Man Who Flew the Memphis Belle," page 5

[3] Astor, "The Mighty Eighth," page 340

[4] Astor, "The Mighty Eighth," page 408

[5] Olson, "Citizens of London," page 257

[6] https://www.imdb.com/title/tt1546368/

[7] https://www.imdb.com/title/tt1546361/

GLOSSARY

10/10 overcast – total overcast

A's – class A dress uniform of shirt, blouse, and tie

A-3 – luggage or duffle bag

AFC – Automatic Flight Control

Air Medal – given to bomber crews after completing five combat missions

APO – Army Post Office

ATC – Air Transport Command

B-4 – luggage or duffle bag

B-17 U.S. Bomber – the aircraft that Daddy flew. For more information, turn to page 203.

B-24 –U.S. Bomber

B-26 –U.S. Bomber

B-29 – Superfortress Bomber used mainly in the Pacific

Ben McLauchlin – Mother's first cousin

Blouse – waist length jacket
BOQ – Bachelor's Officer Quarters
Buzz bombs – early cruise missiles

C-46 – U.S. military transport aircraft
C-47 – U.S. military transport aircraft
C-54 – U.S. military transport aircraft
C-rations – combat rations
Chaff – foil strips released to confuse the enemy radar
CO – Commanding Officer
Columbia Club – enlisted men's Red Cross club
Commandos – prostitutes (see Piccadilly Commandos)

D-Day – June 6th, the date of the Allied invasion of France
DFC – Distinguished Flying Cross, received after 35 missions
Deopham Green – name of the airfield where Daddy was stationed
Duty pilot – designated officer who handles non-flying duties

E-boat – Western Allies' designation for the fast attack craft of the Kriegsmarine; E-boat could refer to a patrol craft. an armed motorboat, or a large torpedo boat.
E-boatyards – where German torpedo boats were produced
EM – enlisted men
ETO – European Theater of Operations
Eggs - bombs
Elizabeth – Daddy's sister
Emmett Clonts – Daddy's friend from Stetson University

Feather – to lessen the air resistance and reduce the drag of a non-functioning propeller
Flak – antiaircraft fire
Flak Farm – a place for combat weary airmen to receive rest and relaxation
Flak happy – mentally affected negatively by flak

Flubbing the dub – to loaf, evade or bungle a duty
Fort – nickname for a B-17 – short for Flying Fortress
Frank Farnbach – also from McIntosh

G2 – military intelligence
Gee Box – system to improve navigation and bombing accuracy
GP – general purpose
Greenhorns – inexperienced people
GST – Greenwich Standard Time

Halifax – British 4-engine heavy bomber
Harold Hatchett – Daddy's best friend, killed in action in Italy
Harry Terrell – Daddy's friend from Ocala, Florida
Heavies – bombers capable of carrying a heavy bomb load with the capability of a long range

IP – Initial point about 20 minutes from the target

Jack Bateman – also from McIntosh
Jerry – slang for German

Lancasters – British 4-engine strategic bombers
LI – light truck
Link – flight simulation trainer

Mae West – life jacket
Marshalling yard – where trains are assembled
ME 109 – German fighter
Mess – meals
Milk run – easy mission
MP – Military Police

P-38 – U.S. fighter plane
P-47 – U.S. fighter plane
P-51 – U.S. fighter plane

Paddlefeet – the non-flying members of the group
Piccadilly Commandos – prostitutes
Pinks – pinkish gray trousers
PFF – Pathfinder Force – an early radar system
PX – Post Exchange

QM – Quarter Master

RAF – Royal Air Force
Rally Point – where the squadron re-formed into combat wing formation
Reindeer Club – officers' hotel and club ran by the Red Cross
Revetment – parking area for aircraft

S-2 – intelligence section
Salvo – simultaneous discharge of bombs
Sam, Jr. – Daddy's brother
Shavetail – newly commissioned officer
Short Arm – visual inspection of the genital area
Slit trench – narrow trench for protection like a foxhole
Slow Time – to allow a new or rebuilt engine a "break-in" period
Stand-down – temporary halt to offensive military action
Stetson University – the DeLand, Florida university from which Daddy graduated
Strafing – attacking a target with an aircraft's machine guns
Swink – also from McIntosh

TO – Transport Officer

Unfolded – moved out of formation

Wilshire – Daddy's older brother
Windmilling – to lessen the air resistance and reduce the drag when an engine is inoperable

SNAKE'S DAUGHTER

About the Author

Helen grew up in her father's hometown of McIntosh, Florida. After graduating from P.K. Yonge Developmental Research School at the University of Florida in Gainesville, she earned a degree in advertising and public relations from Florida State University in Tallahassee and then moved to Winter Park where she married, had daughter Becky and son Joseph.

Following her divorce, Helen worked in residential real estate while raising her children, and then married a most remarkable man, Bob Cairns. Helen and Bob have been together for 28 years, and relish their full and fun-filled life.

The author, Helen Walkup Cairns, and her husband, Bob Cairns, at their home in Orlando, Florida.